Daniel Lirebo Sokido

Density in Relation to Built-Forms and Spatial Quality of Urban Spaces

Daniel Lirebo Sokido

Density in Relation to Built-Forms and Spatial Quality of Urban Spaces

LAP LAMBERT Academic Publishing

Impressum / Imprint

Bibliografische Information der Deutschen Nationalbibliothek: Die Deutsche Nationalbibliothek verzeichnet diese Publikation in der Deutschen Nationalbibliografie; detaillierte bibliografische Daten sind im Internet über http://dnb.d-nb.de abrufbar.
Alle in diesem Buch genannten Marken und Produktnamen unterliegen warenzeichen-, marken- oder patentrechtlichem Schutz bzw. sind Warenzeichen oder eingetragene Warenzeichen der jeweiligen Inhaber. Die Wiedergabe von Marken, Produktnamen, Gebrauchsnamen, Handelsnamen, Warenbezeichnungen u.s.w. in diesem Werk berechtigt auch ohne besondere Kennzeichnung nicht zu der Annahme, dass solche Namen im Sinne der Warenzeichen- und Markenschutzgesetzgebung als frei zu betrachten wären und daher von jedermann benutzt werden dürften.

Bibliographic information published by the Deutsche Nationalbibliothek: The Deutsche Nationalbibliothek lists this publication in the Deutsche Nationalbibliografie; detailed bibliographic data are available in the Internet at http://dnb.d-nb.de.
Any brand names and product names mentioned in this book are subject to trademark, brand or patent protection and are trademarks or registered trademarks of their respective holders. The use of brand names, product names, common names, trade names, product descriptions etc. even without a particular marking in this works is in no way to be construed to mean that such names may be regarded as unrestricted in respect of trademark and brand protection legislation and could thus be used by anyone.

Coverbild / Cover image: www.ingimage.com

Verlag / Publisher:
LAP LAMBERT Academic Publishing
ist ein Imprint der / is a trademark of
OmniScriptum GmbH & Co. KG
Heinrich-Böcking-Str. 6-8, 66121 Saarbrücken, Deutschland / Germany
Email: info@lap-publishing.com

Herstellung: siehe letzte Seite /
Printed at: see last page
ISBN: 978-3-659-61993-9

Copyright © 2014 OmniScriptum GmbH & Co. KG
Alle Rechte vorbehalten. / All rights reserved. Saarbrücken 2014

ABSTRACT

Density is a critical typology in determining sustainable *urban built-form patterns*. Built-form refers to the assemblage and arrangement of the building masses in a city reflecting the spatial layout of spaces. The relationship between density and urban character is also based on at certain densities *(thresholds)*. In a wider sense, sustainable cities are a *matter of density*. Recent debates about the creation of more sustainable urban form, compact cities have led to a renewed focus on issues of density. The argument is that *high density high-rise* with *low ground coverage* or compact city form can offer a high quality of life while minimizing crime, possibility to provide adequate amenities, green and open spaces, accessibility and space consumption. However, the relationships of density and patterns built-form are not reasonably well developed and integrated into the urban planning processes in developing countries cities like Addis Ababa. Similarly, *high density high-rise* dwelling forms with low ground coverage are believed to combat incidence of crime through *'natural surveillance'* or *'many eyes on the street'* like multi-family Condominiums and apartments. On the hand, not many studies have been undertaken with regard to *density in relation to patterns of built-form*s.

Therefore, the study aims to explore *density* in relation to patterns of built-form in the process of planning spatial quality of urban spaces as well as impacts on safety and security of the built-environments. The analysis is based on the primary sources as well as secondary documents collected from the concerned agencies and related references. The findings illustrate the stark realities presented within cities, addressing the ill planned situations of density in relation to patterns of built-form. It concludes with an overview of emerging thinking/implications where further efforts are required in the future.

Key words: Density, ground coverage, floor area ratio, Built-Forms, safety and security, High-Density high-rise and low-rise, Spatial-Quality, crime, built-environment, Addis-Ababa.

ACKNOWLEDGEMENT

Foremost, I would like to acknowledge the input of my advisor and guide. It was a privilege to tap into *Prof. Dr. Sanjukkta Bhaduri's* extensive experience spanning from the first day of her reception as student in her HOD office, School of Planning and Architecture, New Delhi, India to the present. Notwithstanding her long years of experience her perpetual curiosity, symptomatic of the inquisitive "eternal student", was indeed infectious. Influenced by this I have dared to venture out of the comfort zone of my discipline and probe into others turf, striving for a holistic understanding of the subject at hand. Prof. Dr. Sanjukkta's guidance, beyond her decisive comments, had spanned from reading draft chapters to editing important details, always demanding a standard of the highest order. I will remain indebted to her very close guidance, tremendous patience and encouragement; she, true to the spirit of the enthusiastic mentor, so lovingly bestowed. *Prof. Dr. Sanjukkta's* method-oriented advice was also crucial in the struggle of clearing much of the thick fog that surrounds the complexity of such studies. Her thoughtful input was instrumental in pressing the wrinkles that hopefully brought some sense to the messy nature of case studies. Her essential contribution was not limited to guiding me towards achieving a coherent study, but included the reading of all "zero" drafts and language editing. I am also indeed indebted for all these and for her constant reminder of the need to stay on track and attain clarity.

This study is made possible by the financial support of the *Indian Government/ICCR/* and I am indebted to Royal Indian Government. I also extend my heartfelt gratitude to *Prof. Chetan Vidya*, Director of School of Planning and Architecture for encouraging me during the entire period of my stay in the school of Planning and Architecture, New Delhi, India. I would like to thank *Prof. Dr. P.S.N. Rao*, PhD Study co-coordinator, for his concern and support.

During my fieldwork in Addis Ababa, I have had a number of discussions with *Dr. Hailemichael Abera and Dr. Samson Kassahun*, President and academic Vice president of Ethiopian Civil Service University respectively. I am very grateful for their valuable contribution that helped in their unreserved willingness in availing facilities. Finally, I would like to thank the staff members of the Department of Urban Planning, SPA, New Delhi, India, for creating a nurturing atmosphere throughout my stay in the Department.

Daniel Lirebo Sokido
Research Scholar
Department of Urban Planning
School of Planning and Architecture, New Delhi, India

Table of Contents

ABSTRACT---i
ACKNOWLEDGEMENTS---ii
LIST OF TABLES AND FIGURES---iii
ABBREVIATIONS---iv

CHAPTER 1: INTRODUCTION
1.1. Background---2
1.2. The Need of Study---3
1.3. Aim and Objectives---5
1.4. Scope and Limitation of the study---5
1.5. Materials and Method---5

CHAPTER 2: THE CONCEPTUAL FRAMEWORK
2.1. Introduction to the Framework---7
2.2. Key Concepts of the Theoretical Frameworks---8
    2.2.1. Determinants of Patterns of Built-form and Spatial Quality---9
    2.2.2. The Concept of Spatial Quality of Urban space---15
    2.2.3. Spatial quality and Usability of Space in Built Environment---19
2.3. Density and its implications on Spatial Quality---22
    2.3.1. Non-Built Space and Built Total Floor area Ratio---27
2.4. Safety & Crime Prevention in High Density high-rise Housing Forms---28
    2.4.1. Dilemma of High Density and Overcrowdness along with Spatial Quality---32

CHAPTER 3: ANALYSIS & DISCUSSION
3.1. Introduction---34
3.2. Determinants of Built-Form Patterns and impacts on Spatial Quality---34
    3.2.1. Building height---34
    3.2.2. Plot characteristics---36
    3.2.3. Density characteristics and its Impacts on built-forms---37
3.3. The Effects of Floor area Ratio & land coverage at Block & Neighbourhood level---39
3.4. The impact of Built-Up densities on Patterns of Built-form---40
3.5. The Relationship between Population Density & Built-Form Patterns---41
3.6. Safety and Security in Higher Density Built Environment---43
    3.6.1. Higher Density and Crime reduction---45
    3.6.2. The Role of Higher Density on Safety and Environmental quality---46
3.7. Results of Empirical Analysis in relation Spatial Quality---49
    3.7.1. Statistical Model Used to evaluate the results in relation to Security---49
    3.7.2. Model Evaluation with respect to Null and Alternative Hypothesis---53
    3.7.3. Results of Alternative Hypothesis Testing---55

CHPTER 4: KEY FINDINGS AND CONCLUSION
4.1. Key findings---59
4.2. Conclusion---61
REFERENCES

LIST OF FIGURES AND TABLES

LIST OF FIGURES

Fig 1.1 Poor spatial quality and high building coverage

Fig 1.2 Illustration of Built-Up Density in Built-Form

Fig 2.1 Density in terms of Prescriptive and Descriptive

Fig 2.2 Conceptual and operational componentsv of Determinants of Built-forms and Quality

Fig 2.3 Figure 2.3: Conceptual diagram showing the Density in relation to patterns of built-forms and Relationship b/n spatial quality and density in the Built Environment

Fig. 2.4: Illustration OF Different built-forms with the same dwelling density from high rise to low rise, Vicky, 2010: 10

Fig. 2.5: Density and differen t patterns of Built-forms World cities, 2014 (compiled by Author)

Fig 2.6 Plot area, plot ratio and plot exposure respectively

Fig 2.7 Green and open spaces around the buildings, PPS, 2010

Fig 2.8 Example of medium dwelling demnsity in terms of net and gross density, city of Boulder, 2012, USA

Fig 2.9 Conceptual model for FAR values and Ground coverage

Fig 2.10 Built-up densities in different patterns of built-forms

Fig 2.11 Figure-ground diagrams of Collage city, by Rowe & Koetter, 1978:65

Fig 2.12 Showing high desnity high-rise with low ground coverage

Fig 3.1 Building Height in four different case study settlements

Fig 3.2 Plot Exposures in All directions (Lideta-Firdbet) facilitates for better ventilation

Fig 3.3 Plot exposures in four case study settlements

Fig 3.4 Ground coverages in four different case study areas

Fig 3.5 Total floor area ratios in four case study settlements

Fig 3.6 Built-up Density with different patterns of built-forms at plot level and different ground coverage

Fig. 3.7 High density low-rise with high BAR built-form, crowded and poor spatial quality, ventilation by penetrating roofs. No space for circulation (compiled by Author, 2014)

Fig. 3.8 High Density high-rise with low BAR built-form pattern, comprises better spatial quality elements, morphologically good space layout (by author, 2014)

Fig. 3.9 Low Density Low-rise with extreme low BAR, very sparsely settled and residents suffering from incidence of crime, frequent robbery and burglary (by author, 2014)

Fig. 3.10 High density high-rise with High BAR built-form pattern, no adequate spaces for amenities, parking etc. (Gerji-sunshine), (by author, 2014)

Fig. 3.11 The effect of built-form patterns at plot level is being reflected at neighborhood level, or high BAR at plot level can reflected at neighborhood scale (by author, 2014)

Fig. 3.12 Morphology of the four case study settlements with their respective BAR & FAR, showing each case study areas pattern of built-forms (compiled by Author, 2014)

*Fig. 3.13 Different patterns of Built-forms with the same population density, but **high-density high-rise with low BAR** built-form would offer better amenities and spatial quality elements (compiled by author, 2014)*

LIST OF TABLES

Table 2.1: Theories of defining spatial qualities of urban spaces by different scholars

Table 2.2 Dwelling density and built-up density/BAR, FAR/ in different cities   at

neighbourhood scale
Table 2.3 Correlation between density, patterns of built-forms and spatial qquality indicators and their implications
Table 2.4 Space and Approaches to Crime Prevention-Situational Approaches
Table 3.1 Random Sampling (20% of the Total Households in each Neighborhood)
Table 3.2 Test parameters for the binary logistic model not in equation
Table 3.3 Test parameters for the binary logistic model in equation

ABBREVIATIONS

AA--------------Addis Ababa
AACA---------Addis Ababa City Administration
AAU-----------Addis Ababa University
AU-------------African Union
BA-------------Built-up area
BAR-----------Built/Base Area Ratio equivalent to Ground Coverage Ratio
BH-------------Building height
CPTED-------Crime Prevention through Environmental Design
CPTPD-------Crime Prevention through Planning and Design
CSA-----------Central Statistical Agency
DU------------Dwelling Units
Ha-------------Hectares
EC ------------Ethiopian Calendar
ECSU---------Ethiopian Civil Service University
FAR-----------Floor Area Ratio
FSD ----------Focus group discussion
FSI------------Floor Space Index
GS-------------Gerji-Sunshine
GSI------------Ground Space Index
HH------------Households
LDP -----------Local development plan
NUPI---------National Urban Planning Institute
OP-------------Open and Green Spaces
OPR-----------Open Space/Non-Built Space Ratio
ORAAMP----Office for the Revision of the Addis Ababa Master Plan
PA-------------Plot area
PAR-----------Plot area Ratio
SPA------------School of Planning and Architecture
UN------------United Nation
UNDP--------United Nations Development Programme
UN-ECA-----United nation for Economic Commission for Africa
UNCHS------United Nation Center for Human Settlement
UNESCO----United Nations Educational, Scientific and Cultural Organization
VMT----------Vehicle mile Traveleds

This page has been intentionally left blank.

CHAPTER 1: INTRODUCTION
1.1. Background

The city has two main facets, a large collection of buildings 'built-forms' linked by space, and a complex system of human activity linked by interaction. One of the pressing challenges in urban planning processes is controlling these linkages and interactions of the physical structures, spaces and human activities in a given area of built environment through the proper application of *"density"*. *(Van K. & Leduc, 2008:18) state that* the concept of density in urbanism is frequently used to describe the relationship between a given area and the number of certain entities in that area. These entities might be people, dwellings, services, or floor space. However, the simple fact that density is used in, for instance, design requirements, plan descriptions and communication between parties, does not mean that it is used correctly or to its full potential. There are very few efforts being applied by urban planning institutions or professionals and politicians to examine, evaluate and control *densities like built-up densities/BAR, FAR/, patterns of built-forms, population density* and their impacts on the spatial qualities of housing settlements particularly spaces for *walking, standing, sitting, enjoyment and protection as spatial quality elements* in large cities that has to be investigated. Many literature sources also reveal the absence of explicitly defined theories in density and spatial qualities.

It is strongly believed that urban planning is in a crisis particularly due to the absence of clear-cut theories of spatial quality and density of built environments. Planners and politicians over the world are aware of the urgent need for action plans to increase the spatial quality and sustainability in the large cities like Addis Ababa. The campaign for improved urban affects a broad field of action: influencing the attitudes and lifestyles of the inhabitants, developing alternative access amenities and facilities, safety and security, green and open spaces, as well as changing the urban structure, the land use pattern etc. It is clearly a formidable task, both in terms of time and money needed. But greatest obstacle is probably the lack of an adequate theory in relation to the issues of *spatial quality and density paradox*. Everybody will agree on the need to plan for the sustainable city. For instance, Green Paper on the *Urban Environment*, 1990), what would the ideal sustainable city look like? (Rådberg, 1996) underlines that there is no definite answer today. On one hand, we have the proponents of the compact city. On the other hand, we have the proponents of the green city. The first implies a strategy of concentration and increasing urban density. The second implies a strategy of deconcentration and spreading out, using the unbuilt land. So we are confronted with two strategies seemingly irreconcilable for the sustainable city as the "density paradox".

Therefore, there is a great confusion in the fields of *spatial quality and sustainability* in urban planning. Much of this confusion stems from the fact that the theories are formulated on a very general and abstract level. We need empirical facts, observations. Above all, we need a theoretical framework for these empirical observations. We need a systematic descriptive classification of the urban structure on the micro-level in order to be able to process the accumulated information on existing urban spaces of built-environments. This study will focus on the issues of urban densities of built environment, patterns of built-forms and spatial qualities of housing settlements to come up with clear cut theories along built environments.

1.2. The Need of Study

We are in the urbanising world. Aside from the growth or urbanisation itself, urbanisation is the dominant demographic trend of our time. 150 million living in cities 1900 swelled to 3 billion more by 2007, a 20 fold increase. It is also strongly believed that more than 70% of world population will live in cities and towns by 2050 (UN, 2007). One of the most pressing challenges of 21st century urbanization is the problem of providing public amenities, green and open spaces and other communal facilities for the rapidly growing urban population and the problems connected to the application of *"density"* in the built-environment. *(P.H.M. Steven, 1960:6) Wisely used, density can be a valuable weapon in the planners' 'armoury', but indiscriminate use has revealed some limitations.* And hence, *the problem of density* should be resolved *by density* through appropriate application of planning and design in the built-environment so as to install good quality housing settlements in the city.

The practice of urban planning is argued to be unsuccessful in creating a healthy urban environment both in terms of its methods and effective *'use of **built and unbuilt** urban spaces'* and the application of *'density'* during the preparation of city plan and its implementation process *(see figure 1.1.)*. This situation causes a handicap for the creation of quantitative/qualitative quality living spaces as one of the main concerns of urban planning discipline. It contributes to the improper uses of built and or unbuilt spaces, formation of low population density, low floor area ratio/FAR/ and large scale absorption of land/plot by high building coverage/BAR/. A naked eye observation on current urban morphologies in many large cities in the world, especially in the cities of developing countries like Addis Ababa suffices to retrace the minimal role of planning to cope-up the challenges as the result of urbanisation and urban development. This predominance of haphazard development in the 21st century urbanisation poses a substantial need for re-planning and re-designing cities of the World to use limited resource efficiently by setting density thresholds to develop quality urban spaces & sustainable built-environment.

Therefore, the norms of "urban space consumption" should have to get emphasis through the proper planning and design of *built-up density /BAR, FAR/* and dwelling density with proper **built-form patterns** to install better spatial quality of housing settlements. In this study, **urban space refers** to both *built (residential, mixed use building structures) and non-built (Green, open space, spaces around buildings and outdoor spaces, parking, circulation/streets & etc) spaces* morphologically as their (Built & unbuilt space) strong integration offers good spatial quality of urban spaces for **walking, standing, sitting, enjoyment and protection,** like possibility to use outdoor spaces, public amenities, adequate green and open spaces, maximum ventilation, access to daylight, less or no incidences of crime, short travel distance and etc. Hence, **spatial quality of housing settlement** is defined as the effectiveness and capacity of urban spaces to function the communities properly with higher performance efficiency of space usage. Quality of space-usage is the clear manifestation of strong integration and harmony of built and non-built spaces or integrated *solid-void* relationship in the built environments of the city.

*However, literature reviewed reveals that built environ*ments in many large cities of the World, particularly in Africa, some parts of Asia and Latin America are highly composed of single storied houses/low FAR/ and mostly dense due to high ground coverage/BAR/ or the close distance between houses as improper uses of spaces inside the blocks hinder the spatial quality in the housing settlement. The usability of spaces inside the blocks is not efficient because most of the spaces inside the blocks and plots are covered by buildings. In order to increase the spatial quality of urban space, there is a need to analyse the built-up density in terms of floor area ratio (FAR) and percentage of ground coverage/BAR/ along with different *patterns of built-forms*. Hence, FAR means the ratio between the total floor area and the total land area. BAR means the ratio between built up area and area of plot or block in the settlement.

This study analyses spatial quality of housing settlement in terms of provision of green and open spaces, mobility and circulation, accessibility, less or no incidence of crime, possibility to use outdoor space inside the plots and or blocks, optimum ventilation, and access to daylight and the likes. Especially, in cities located within hot, hot humid and semi-humid climates, the dwelling environments have been organized to facilitate maximum cross-ventilation and comfortable use of spaces. In a different context but similar climate situation *khan* notes that:

"For people in the tropics, it is normal to eat, work and play outdoors and to seek shelter of a house only when the need for privacy demands it, and thus the treatment of the adjoining ground should be seen as an extension of the homes, Outdoor life in a warm humid climate is only pleasant if there is a breeze, shade and protection from rain" (Khan, 1985:25).

On the other hand, one of the pressing challenges in the urban development process, particularly in developing countries cities is the **"gap of knowledge"** associated with *"concept and theory of densities and spatial qualities"* as well as their impact on *"quality of urban space"*. As it has been argued that inadequate knowledge base on *patterns of built-forms, prevailing dwelling and built-up densities/BAR, FAR/*, urban space usability and plot characteristics that take place in the built environments have *"restricted adoption of effective planning"* inventions. With this regard, Nnkya addresses tha*t:*

Fig.1.1: Poor Spatial Quality and High Building Coverage/BAR/

*"the lack of, or too little knowledge, on the social, economic and political processes, which shapes the physical environment has been influential to **"defective planning"** and in some instances triggered off disputes between the planning authorities and the stakeholders"* (Nnkya, 1999:19).

Accordingly, the study helps at improving the understandings among the scholars who are supposed to generate methods, guidelines and theories about the density and spatial quality of housing settlement in the built environment so as to narrow the knowledge gap in the fields of *physical and spatial planning.*

Therefore, this study aims to investigate the relationships between *"density and spatial quality"* that would show the impact of dwelling and built-up density ((FAR/BAR)(*see figures 1.1 and 1.2*)) of built-environments on spatial quality to keep pace with the expanding horizon of knowledge that provides strong theoretical grounding looks for application of knowledge to develop solutions. It can also help the policy makers, managers, urban planners and city officials to improve their understandings on the issues of spatial quality and density. On the basis of these realities, the following problem statement, hypothesis and objectives have been stated for investigation.

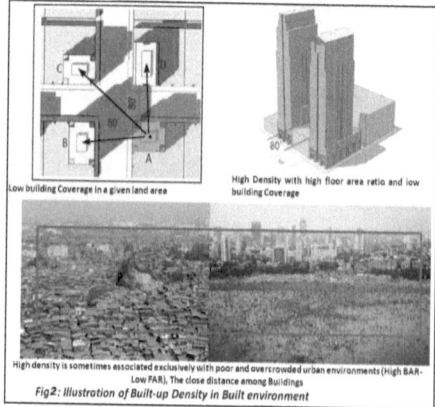

Fig.1.2: Illustration of Built-Up Density in Built-Form

1.3. Aims and Objectives of the Research

"It is with the above hypothesis, the aim of the study is to analyse and explore the relationship between densities and spatial quality of the settlements/neighborhoods of the city so as to contribute to a better understanding of the correlation between density and spatial quality of housing settlements". Therefore, the urban density particularly *population and Built-up Density (edificatory) (Floor area Ratio/FAR/ and Ground coverage Ratio/BAR/)* along with various patterns of built-forms *and or* housing forms including *high density high-rise with low ground coverage, high density low-rise and high-rise with high ground coverage,* and *low density low-rise with extreme low ground coverage* by housing structures within different case study neighborhoods are studied if they are directly or indirectly affecting the elements of the spatial qualities of urban spaces in the housing settlements, the case of Addis Ababa, Ethiopia.

1.4. Scope and Limitation of The Research

The scope of the study is limited **Addis Ababa, Ethiopia**. It explores and investigates the relationships between urban densities, and spatial qualities in **within four selected Neighborhoods'** specifically deals with **population and Built-up Density, Spatial quality**. The main focus was on the urban density **(Population and Built-up Density/BAR, FAR/) & Physical Quality,** Building Heights, set-backs, Incidence of Crime, urban built-form, safety and security, ventilation and daylight access, and possibility to use outdoor spaces. The study engaged *Stage-I: (13 neighborhoods rapid appraisal of the entire city) to select four neighborhoods for detail study, Stage-II: Detail Data collection from the selected four neighborhoods (four neighborhoods for Detail study)* with different patterns built-form & density characteristics including: *High BAR vs. Low FAR or High Density Low-rise with high Ground*

Coverage, High BAR vs. High FAR or High Density High-rise with High Ground Coverage, Low BAR vs. High FAR or High Density High-rise with Low ground Coverage and Low BAR vs. Low FAR or Low Density Low-rise with extreme Low Ground Coverage.

Therefore, the impacts of *urban densities on spatial qualities* of different neighborhoods were studied if they were directly or indirectly affecting the spatial quality of the housing settlements in Addis Ababa, Ethiopia. Although the blocks of urban spaces from city center and the surrounding were studied, the main focus was on the urban density (built-up density with different patterns of built forms, population density), Floor area Ratio/FAR/, Ground Coverage Ratio/BAR/, Building Heights & set-backs, connectivity, *safety and Security*, space usability, optimum ventilation, and possibility to use outdoor spaces in the selected case study settlements. The study has been engaged *four* settlements from city centers, intermediate & suburbs for investigation, *2 from (Inner city), 1 from intermediate zone* and *1 from (sub-zone-suburbs)* of the city. Limitations of this study were financial constraint and shortage of few relevant data. On the other hand, the willingness of some inhabitants to respond the questions and interviews were also happening in the study area during data collection.

1.5. Materials and Methods

This study has been employed the case study approach. The case study approach advocates the use of multiple sources of data and data collection methods, (Yin, 1981 in Kombe, 1995:55). The approach to data collection and analysis in this study included both *quantitative and qualitative* sources and approaches, *'triangulation'*.

The tools that were used under quantitative method include: household survey (questionnaire) at neighbourhood level, measurements along with support of (Google Earth, GIS/Line/Nortek Maps), (up to 20% of plots and housing units were surveyed from each neighbourhood). Under qualititive method the main tools were visual survey, open ended interviews with local and city level authorities, key informants, and communities; and focus group discussion. The primary data have been augmented by secondary data obtained from documents from planning institutions and concerned agencies during analysis. Furthermore, related books, journals and websites have been consulted.

The study embraced the plots, blocks and neighborhoods from city center to suburb were studied. The area of each case study settlement ranges between *10.2-10.7 hectares* with population density of *154 to 647inh/ha*. Out of these areas of each case study about 20% sample of household and the built environments have been be surveyed. And hence, the study has been engaged *four neighborhoods or* settlements including; *2 from Inner city, 1 from intermediate zone* and *1 from (sub-zone)* of the city. The selection of case study areas was done on the basis of the criteria matrix as per the identified indicators and variables under study. In spatial quality empirical investigation part, binary logistic regression analysis has also been employed.

CHAPTER 2: THE CONCEPTUAL FRAMEWORK
2.1. Introduction to the Framework

How humans have come to *'use urban space'* over time in some cases judged as too intensely, in others not intensely enough and the problems connected to this, have resulted in concerning the application of the concept of *'density'* in urbanism. *Today high densities and the compact city are often seen as prerequisites for sustainable urbanisation and economic growth (van K. & Leduc, 2008:18).* However, density is a rich but unresolved concept in urban theory. There are presently two developments in the process of urbanisation which can be identified that legitimise the study of density. First, recent changes in how city building is organized have created a greater need to relate development programmes to spatial qualities. Second, the trend in the increase in space consumption and the environmental, economic and social effects associated with this trend point to the need for investigation into the relationship between the quality and capacity of space *(M. Berghauser et al., 2010:16).*

Very few efforts are being applied by urban planning institutions or professionals and politicians to examine, evaluate and control building densities and their impacts on the spatial qualities to maintain sustainability in cities. *But greatest obstacle is probably the lack of an adequate theory in relation to the issues of spatial quality and density paradox (Rådberg, 1996:385).* Therefore, it is strongly believed that urban planning is in a crisis particularly due to the absence of clear-cut theories of spatial quality and density of built environments. Planners and politicians over the world are aware of the urgent need for action plans to increase the environmental and or spatial quality and sustainability in the large cities like Addis Ababa, Ethiopia. The campaign for improved urban affects a broad field of action: influencing the attitudes and lifestyles of the inhabitants, developing alternative amenities, communal (outdoor, green and open spaces), mode of transportation systems, as well as changing the urban structure. It is quite evident that the application of density has been misunderstood and not yet clearly addressed. It requires clear distinction in between the density used to describe built environment (**Descriptive**, has legal constraint and only to illustrate built environment) and urban density used as a norm in the process of planning the city (**prescriptive**, normative and has legal status to design built environment) *(figure 2.1).*

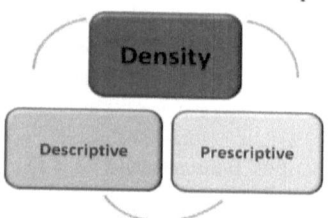

Fig. 2.1: density in terms of Prescriptive and Descriptive

The issue of spatial quality, urban density and urban built-form is no less confused. The leading theorists of urban planning and design during the twentieth century (like, Le Corbusier, R. Unwin, L. Mumford, J. Jacobs, K. Lynch and C. Alexander) argue in favour of different ideal solutions. They argue for different urban models: some advocate **high density high-rise** buildings and large open spaces, others argue for traditional grid-iron plans, streets and compact blocks, still others argue for small-scale garden suburbs. There is evidently no consensus on the question of urban spatial quality, level of density and urban built-form. Therefore, at certain densities (thresholds), the number of people within a given area is sufficient to generate the interactions needed to make certain urban functions or activities viable. Clearly, the greater the number and variety of urban activities, the richer the life of a community; thus, urbanity is based on density' (Lozano, 1990: 316).

The literatures reviewed reveals that there is a great confusion in the fields of *spatial quality and sustainability* in urban planning. Much of this confusion stems from the fact that the theories are formulated on a very general and abstract level, so that it needs empirical facts, observations. Above all, planners/architect-designers/ need a theoretical framework for these empirical observations as well as they need a systematic descriptive classification of the urban structure on the micro level in order to be able to process the accumulated information on existing built-environments. This study stresses on the issues of *density and patterns of built-forms* as determinants of *spatial qualities of urban spaces* in the housing settlements conceptually to come up with clear understandings.

2.2. Key Concepts of the Theoretical Frameworks
2.2.1. Determinants of Built-Form Patterns and spatial Quality

Determinants of built-forms as employed in this study refers to parts of settlements having a common set of physical characteristics in terms of built-forms, density, urban spaces and plot characteristics would exert impact on spatial quality. The conceptual and operational definitions of this urban built-form component in the as applied in this study are summarized in *Figure 2.2 below*.

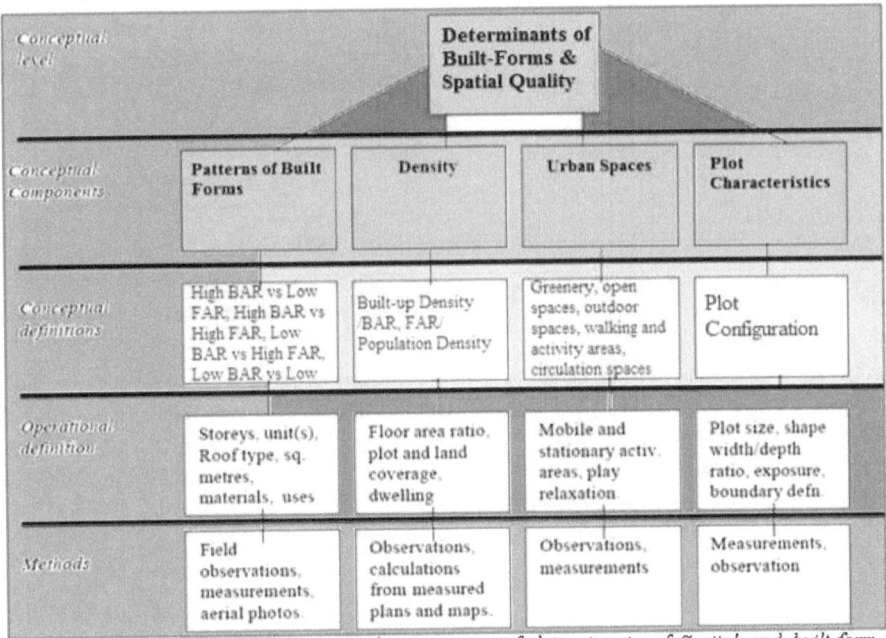

Figure 3.2: Conceptual and operational components of determinants of Spatial and built-form (Source: adapted, Nachmias and Nachmias, (1997: 33) modified to relate to the context of this study).

2.2.1.1. Density in relation to Patterns of Built forms and its Implications

Built form includes the assemblage of buildings and the massing of low and high buildings, which influence the internal external comfort conditions (Sanjukta, 2002:46). The built form in urban area changes with the population growth of an urban area with the change in massing, height etc. due to the changing intensity of land use. Massing influences the pattern of wind movement culminating in temperature changes in the surrounding area. The mosaic of buildings, spacing between tall and low structures, the design of building mass and spaces and surfaces of different thermal characteristics affect the redistribution of solar radiation causing change in the outdoor temperature to achieve and maintain the comfort conditions within the building as quality of urban spaces. 'Once constructed, the built environment constrains the level and pattern of built-form demand for example in low density suburbs, segregated from employment and service opportunities and lacking efficient public transport systems, the inhabitants are necessarily dependant on a high level of personal mobility as a good indicator of poor spatial quality.

It is also important to note that Buildings are designed and constructed firstly for achieving the functional purpose (for which they are built) and secondly for achieving the conditions for comfort within them. The depth of buildings, height, orientation, design, and construction materials and techniques should be carefully considered in order to achieve the desired objective of providing lighting, ventilation and adequate comfort conditions in the interiors and exteriors of buildings to result good spatial quality of built-environment. Consideration of climatic factors in design i.e. need for solar access, ventilation etc. imply selection of appropriate orientation for the building.

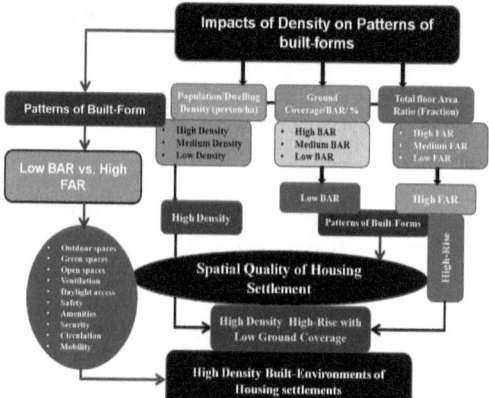

Figure 2.3: Conceptual diagram showing the Density in relation to patterns of built-forms and Relationship b/n spatial quality and density in the Built Environment

'Siting and orientation have important spatial quality implications since they can be used to gain advantage from microclimatic factors.

On the other hand, Karen Franck discusses form in relation to places. She says that there are three attributes of place types namely *form, use* and *meaning* (Franck, 1994:346). While form attributes include all the material, spatial, structural and geometric properties, use attributes comprise all aspects of use which are sometimes referred to as function involving the performance of specific tasks that are housed by a particular type. Meaning attributes comprise the practical and symbolic messages that are conveyed by aspects of form and use or that are more loosely associated with that density and spatial quality (Franck, 1994:346-347).

On the other hand, Moudon argues that in order to characterize built forms, inventory on built forms should aim at observing and documenting dwelling forms in terms of their shapes, the major building elements and where necessary their decorative elements (Moudon, 1994:294). Formal characteristics also include whether houses are detached, semidetached, row, or high-rise buildings. The number of storeys, roof type, building materials, house sizes, and building uses are essential elements that characterize built forms. King denotes that the act of placing houses into either the same or different classes requires the selection of only those features that are seen as significant for making a distinction between them (King, 1994:130).

However, the criteria used to classify and elaborate the degree of differentiation between them are obviously related to the purposes for which the categories are created. King, for example, provides taxonomy of house types in England based on official government statistics as detached houses and bungalows, semidetached houses, terraced houses, flat and maisonettes and other accommodations (King, 1994:131). In this study, built forms are being considered not only as exemplars in the classification of housing typologies but also as analytic variables of the housing settlement. Pertinent questions

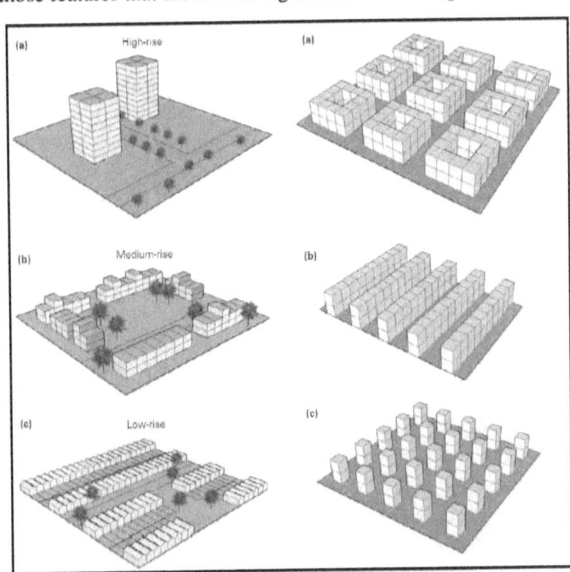

Fig. 2.4: Illustration OF Different built-forms with the same dwelling density from high rise to low rise, Vicky, 2010: 10

that warrant investigation are: What are the dominant built forms prevailing in the residential neighborhoods?? How significant are density and built forms in defining built-environment? What socioeconomic features have influenced the development of these built forms and its implication? What are the patterns of built forms in relation to density? What are the implications of different patterns of built-forms on spatial quality of urban spaces?

Accordingly, pattern of built form is in one of the determinants as illustrated in the conceptual diagram above, which would strongly influence spatial quality of housing settlements *(figure 2.3)*. It is also worth mentioning that by keeping population density constant, we can design built environment with different built forms which might influence spatial quality like *high density high rise with low ground coverage, high density low-rise with high ground coverage and medium density medium-rise with medium ground coverage categories of density* in built-environment *(see figure 2.4)*.

Therefore, with different scenarios of patterns of built-forms, population density can be remaining the same; however the level of spatial quality might be different for each built-form. For instance, *high density high-rise with low ground converge* can offer higher quality urban spaces by creating better possibilities to provide public amenities, green and open spaces, visibility, ventilation, safety and security(*many eyes on the street*) and likes as illustrated on (*figure 2.4*). These figures clearly demonstrate that the same **dwelling density** in different patterns of built-forms as stated above. Each built-form patterns has different characteristics and implications like high density high-rise with low ground coverage has offered possibilities to provide communal outdoor spaces, adequate amenities, safety and security, circulation and mobility and other spatial quality elements in the neighborhood, whereas in low rise with high ground coverage occupied larger spaces by building structure, which has limited possibilities to offer important elements of spatial quality of the neighborhood for the residents at large. This is clear manifestations of how far the different **patterns of built-form** with the same population/dwelling density are affecting spatial quality of the settlement. Therefore, that is why pattern of built form is considered as one of the determinants of spatial quality.

2.2.1.1.1. Built form and Non-built space

Buildings do not exist in isolation; rather they form a part of the cityscape or townscape in conjunction with other buildings and urban public spaces. The size and shape of plots also influence the internal external thermal comfort conditions and subsequently the availability of amenities. Disproportionate length and width ratio of a plot of land especially in case of semi-detached plots causes inappropriate or increased depth of the building. The depth of buildings and the orientation of buildings dictate the interior thermal comfort conditions and lighting requirements through sunlight access and wind flow within the building. The depth of buildings in case of row housing is an important factor for the amenities requirement of the unit. With openings only on the front and rear sides of the building, excessive depth of the unit results in inadequate natural lighting in the interior exterior of the block in the residential neighborhoods.

Not only is it important to consider the comfort, lighting etc. in the interiors of buildings, it is also necessary to analyze the impact of built form on the microclimate to facilitate spatial quality of the housing settlement. Built form and its interrelation with the open space influence the microclimate of the complex, which in turn affect the comfort conditions and resultant space consumption in between building. The mosaic of buildings, the spacing between tall and low structures, building mass, spaces and surfaces of different thermal characteristics affect the redistribution of solar radiation (affecting also the lighting of interiors) causing an increase in the outdoor temperature. Use of mechanised cooling devices to achieve comfort levels i.e. greeneries in the surroundings to balance microclimate condition in the built environment.

2.2.1.2. Patterns of Built-Form and its Influence on Spatial Quality

There has been varying perceptions among scholars and authors to the concept 'density'. These perceptions stem from the varying fields of discipline from which density draws its meaning. Density has been studied extensively from many perspectives including – physical, psychological, social and environmental (Acioly and Davidson; 1996; Rådberg, 1996; Mahbubur, 2001 and Arenas-Gomez,

2002). In the field of planning, a misunderstanding arises because of the several kinds of density used such as population density, built-up density/BAR, FAR/, net density, gross density etc. (James, 1967:552 and Alexander, 1993). Another ambiguity arises from the use of the concept without clearly defining it. Mitrany and Churchman argue that in many studies, density has been referred to as *"high"* or *"low"*, without a definition of what is a high or a low one. As a result these studies have not built up a sufficient body of knowledge or a comprehensive theory about the meaning of residential density (Mitrany and Churchman, undated). This argument has been also raised by Correa who contends that:

*...the old indicators of so many square metres of open space per 1000 persons are too simplistic and crude; we have got to desegregate these numbers, both qualitatively and quantitatively in order to anticipate their real usefulness. Estimating accurately the production costs of these various spaces involves examining the relation between building heights and overall densities, since the latter is the key determinant of **built-forms** at the city scale.* (Correa, 1985:39)

High and low densities are relative measures. They differ between countries and communities, and they are dependent on which perspective density is being discussed. Comparison in density especially perceived density is complicated, since impressions and personal judgments are different. Ernest Alexander argues that there is no simple, clear definition of perceived density. Rather, it is a complex concept involving the interaction of perception with the concrete realities (Alexander, 1993:183). But the central question in this study is how density can be used as an exemplar and as an analytic tool of urban spatial quality and a measure of optimal land utilization.

Density has often been referred to as a degree or intensity of development or of occupancy as patterns of built-form is **an outcome of density** in built-environment. Conventionally, urban densities have been defined from two perspectives; of population and built-up density/BAR, FAR/. While population density has been referred to as the number of persons per unit ground area of development, built-up density (sometimes referred to as objective density) has been examined as land use ratios. In housing and urban design, density has been measured in terms of *floor area ratios/FAR/, ground coverage/BAR/* and *dwelling units per specified area* (Alexander, 1993). Accommodation density in housing has been expressed as the number of inhabitants per unit of habitable space. Floor Area Ratio/FAR/ is a unit of density referring to the floor space in relation to plot or land area. Most of the space standards used for commercial or shopping areas, residential, institutional areas etc. are based on this unit of this type of density. On the other hand, the correlation between density and patterns of built-form is quite strong that they would directly and indirectly influence spatial quality of housing

Fig. 2.5: Density and differen t patterns of Built-forms World cities, 2014 (compiled by Author)

settlements. Hence, density and patterns of built-form are sorts of determinants, which would affect the quality of urban spaces as they can't exist in isolation to each other.

It is also very interesting to note that pattern of built-form is a product of the application of well or badly designed density as density and built-form are different faces of the same coin, which directly or indirectly exert impacts on spatial quality of housing settlements in the built-environment *(see figure 2.5)*. Thus, they are parts of the major determinants, which influence the quality of urban spaces. It is obvious that density could be presented in terms of population density, ground coverage at different level/BAR/ and total floor area ratio/FAR/ that might affect the patterns of built-form as changes in density characteristics changes the patterns of built-form. It is also strongly believed that Density and patterns of built-forms could also be demonstrated in various scenarios including *high density low-rise, high density high-rise, medium density medium-rise, and medium density low-rise, low density low-rise with different levels of ground coverage ratios/*BAR/*in the built-environment as illustrated on figure 2.5 above.*

The review from the above illustration unveils that high density high-rise with low ground coverage built-form patterns offers better spatial quality elements in the neighborhood that the other patterns of built-forms like high density high-rise and low-rise with high ground coverage have limitations to offer spaces for walking, standing, sitting and protection, because major spaces are occupied by building structures that they have been failed to create possibilities to adequate spatial quality elements including communal outdoor spaces, green and open spaces, security and safety, ventilation and the likes as also illustrated on *figure 2.5 above.*

2.2.1.3. Plot configuration and characteristics

The size and shapes of plots play an influential role in the definition of housing forms since they finally determine the amount (*size, ground coverage, floor space ratio/FAR/ and orientation*) of buildings that can be built up. Acioly and Davidson argue that plot size is a culturally bound phenomenon and therefore varies from country to country. Empirical studies show increasing trends in plot size from Asian, Latin and Southern American countries with the largest plot sizes recorded in many African countries (Acioly and Davidson, 1996:16). Further to these observations, Acioly and Davidson discuss the impact of narrow plots with respect to ventilation and ensuing house types and forms. They argue that:

The narrower the plots the more will fit in a particular cluster pattern which is often pre-defined by urban design regulations. Narrow plots impose design and usually imply very narrow houses, narrow rooms with a housing expansion towards the backyard, especially if minimal setbacks to allow circulation, ventilation, light are respected. (Acioly and Davidson, 1996:16-17)

In informal settlements where plot dimensions and size are determined by informal transaction of the land market, formal urban design regulations might not have a significant influence on the ensuing types of plots and houses. Carlos Barquin *et al* argue that the single most important planning decision in designing sites and services projects has probably been the determination of the area of the plot. They argue that plot areas vary widely from region to region, from as small as 18 square metres in India to more than 100 square metres in Africa and South America. There have been little agreements as to what constitute an optimal or even a minimal plot size. Whereas public authorities have responded to the high cost of land and infrastructure justifying the reduction in plot sizes, critics of small plots have pointed out that the larger the plot the greater the economic benefit to the owner and over time, one can improve and invest in his or her home. It has also been suggested that plots in 'slums' and informal settlements are frequently larger than those found in formally planned sites and services reflecting the higher priority that users themselves put on plot size (Barquin et. al. 1986:3).

In rapidly urbanizing cities like Addis Ababa where conventional planning is having limited influence over the sprawling city and of the ensuing housing typologies, a study on the evolutionary nature of plots and plot characteristics is much more demanding in terms of investigating how plot subdivision evolves and in search for optimal plot sizes that will provide a basis for addressing apparent problems of city sprawl. Characterising plot in low-income settlements in India, Carlos Barquin *et al.* employs the variables of plot area, plot ratio and exposure to analyze the structure and development pattern of these settlements. Plot area simply refers to the size of the plots and enables one to assess variations across settlements, while plot ratio refers to the proportion between the width and depth of the plots. Plot ratio indicates the basic shape of the plots. The underlying assumption of plot ratio has been that the more the '*squarish*' shape, the less effective the layout in terms of infrastructure utilization since fewer houses will have access to these infrastructure lines.

Barquin *et al.* analyses plot exposure in the Indian low-income context with relatively small plots sizes ranging from 15 to 37 square metres and observe:

When plots are very small (smallness will depend on market demand, user perceptions and living habits), the plot is entirely built over. This phenomenon occurs in many countries where owner builders inevitably build over space that was intended for patios and courtyards. This is certainly the case over low-income urban settlements in India. In a house that entirely covers the plot, the only possibility for door and windows will open on those sides of the plot that are adjacent to unbuilt public open space. Hence the importance of exposure is as a measure of amenity. (Barquin et al., 1986:7)

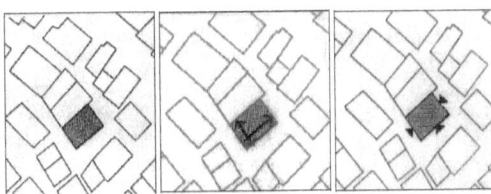

Fig. 2.6: Plot area, plot ratio and plot exposure respectively, (Source: Carlos BaRquin et al., 1986;2)

Plot exposure refers to the number of sides the plot or building has that are contiguous to public open spaces or streets. In hot semi-humid climates like Addis Ababa and in settlements where plot coverage is so high, plot exposure facilitates cross ventilation, sunlight, view and where applicable outdoor space for outdoor activities or living, however, the situation of plot exposure is highly abused in African cities like Addis Ababa due to extreme ground coverage by housing structures. Plot exposure as employed in this analysis explores this variable as a measure of amenity. As far as analysis of plot configuration and characteristics are concerned, a number of questions bear examination. How do plots evolve and get transformed over time and exert impacts on spatial quality? What are the variations of plot sizes in housing settlements? How do these variations influence housing forms and spatial qualities? Are plot sizes and plot ratios consistent with ideas of optimal land and space utilization to make sure spatial quality? *(see figure 2.6)*

2.2.2. The Concept of Spatial Quality of Urban space

Before discussing the concept of spatial quality of urban spaces, it is very important to realize that urban spaces should be clearly addressed. Thus, according to Krier (1979), urban spaces are defined as all types of spaces between buildings and geometrically bounded by elevations. Similarly Spreiregen (1972) defines urban spaces as formal spaces which are the products of cities and usually moulded by building facades and the *city floor*. He also indicates that these spaces must be distinguished from other spaces by their predominant characteristics such as their quality of enclosure; the quality of their detailed treatment or outfitting; and the activity that occur in them. He added that if anyone of these qualities is sufficiently strong, it alone may establish the *sense of urban space and place* as well. In discussing urban space, Carmona also makes a useful distinction of urban spaces as: "hard space" principally bounded by Architectural walls, and the "soft space" or parks, gardens and linear greenways which have less enclosure or defined boundary and are dominated by the natural environment. He further elaborates, for all hard urban spaces three major space defining elements exist: *these are the surrounding structures; the floor; and the imaginary sphere of the sky overhead* (Carmona et. al., 2003:138).

Figure 2.7: Greenery and Open space Good example

Source: Greenery and open spaces arround the buildings, urban space project, 2010

Paul Spreiregen also describes the size variation of urban spaces an: *"Urban and architectural spaces from a hierarchy of spatial types, based on their size. In urban planning and design this hierarchy ranges from the scale of small intimate court spaces on to grand urban space and culminating in the vast space of nature in which the city is set"* (Spreiregen, 1972:126). In relation to urban spaces, many practitioners on the other hand have been attempted to investigate their qualities by defining urban space spatially to what extent they are serving the communities at large in built environments *(see fig.2.7)*. Accordingly, (M. Goethals, 2007:4) defines spatial quality of urban space as the extent to which that space satisfies the expectations of a community. These expectations are determined by the values pursued by the community for its development, more specifically its spatial development. They are expressed both in very general terms, the values pursued by the community and in very specific configuration principles for that space.

This clearly addresses that he emphasizes both the specific configuration principles and the general terms with which a group of persons express their expectations regarding the space as terms for defining the quality of urban space. For this purpose, it is based on the definition of *spatial configuration concepts*. Although spatial developments also include changes in the management and use of space, very often physical interventions are also carried out. A configuration concept has consisted of a configuration proposal and the underlying concepts concerning quality, the spatial situation and feasibility. It has been argued that there is no explicit definition and clear cut theory for spatial quality of urban spaces in many literatures. However, some concepts that might be incorporated in the description of spatial quality were identified. These concepts include ideas about "good city form" (Lynch, 1984), "urban quality" (Trip, 2007), "effective planning process" (Conroy and Berke, 2004), "good design" (Sternberg, 2000), "quality planning" (Creedy et al., 2007), "place quality" (Healey, 2004), "spatial justice" (Soja, 2010), and "fulfilments of human needs" (Moulaert, 2009) in the different readings *(see table 2.1 below)*.

In the series of theories on urban design principles (like Camillo Sitte, Gordon Cullen, Christopher Alexander, Kevin Lynch) the objectives to be pursued for good environment are expressed not only in terms of proposals for new urban structures, urban fabrics, the planning and design of public spaces, accessibility, but also in more general terms and checklists for good urban planning. Whereas Cullen and Sitte place the main emphasis on the visual perception of space `*the city with its artistic principle*`, '*A Theory of Good City Form*' (Lynch, 1981) approaches the perception of space from different angles. In so doing, this theory connects best with Sternberg's description of perception of spatial quality as *"good design"*. He interprets perception more broadly than purely sensory and aesthetic perception. Perception is also a matter of experiencing how the space functions and is appreciated on the basis of a cultural value pattern, the degree to which the form of the built environment influences its perception in a positive way through appropriate application of '*density*'. Perception at the same time means sensory observation and understanding. The inhabitants or users will identify more readily with an environment that supports observation well.

The understanding and appreciation of built environment is strongly determined by a person's knowledge or framework of reference with regard to urban spaces. Accordingly, more recent theories concerning the definition of the quality of space (Oswald & Bacini, 2003; Carmona, 2004) likewise approach space from different angles. In contrast to Lynch's approach, these more recent theories are based on a specific urban paradigm. Carmona sets out, for sustainable urban planning and design, Oswald & Bacini from the modern network city. Lynch's theory can be applied to various urban models. Although his '*dimensions of performance*' are directly related to urban space, they do not impose any specific planning and design solutions.

From the above dimension of performance for spatial quality of urban spaces, the author deeply understands that how sense of place develops and changes is relevant to understanding how people interact with their built environment in general and considering how this interaction may become more sustainable, while the link between density and spatial quality of urban spaces would have got strong. This could also be explained as living quality, the result of millions of individual acts of transformation to create fitness, is what gives a place its placeless and quality urban space. But in order for this quality to emerge, there must be a personally enabling urban process at work. It has been argued that Lynch's *sense of place* is a general term for the recognition, every community and landscape is unique and deserves to be treated with that urban space for **walking, standing sitting and protection** as indicators of spatial qualities of urban spaces. Community planning is sometimes called place making. A sense of place means that people care about a community or landscape, and that policy and planning decisions reflect their concerns. When people work to improve their existing community or protect a unique landscape feature, greenery and soft landscaping, they are showing a sense of place by creating quality urban spaces in built environment.

The literature that reviewed leads to the conclusion those terms for defining spatial quality can equally well be developed on the basis of a specific urban model and on the basis of rather general expectations that people have of space. Lynch's framework of analysis offers the most inclusive view on spatial quality. This framework is not confined to one singular city model. It is suited to examine multiple models. On the hand, (K. lynch, 1981) in his book *"The Image of the city"*, also underlined five basic elements of image of the city including **Node, edge, path, land mark and district** which are used to recognize the *spatial quality* of urban spaces. *Furthermore Lynch offers two important views:* the meta-criteria *'Efficiency'* and *'Justice'* enable to weigh Lynch's dimensions of spatial quality against forces that define space, but without a conscious intention of creating spatial quality. The covering of meta-criteria provides the possibility to relate spatial intentions to more global social intentions, such as democracy and pluralism, for all social units of the community.

The below, all authors, including Lynch offer the most elaborate analysis of the spatial perception. On the other hand, Carmona also stressed the *diversity* and *choice* as principles of spatial quality so as to make sure sustainability. Author stresses this idea, because all social communities should have to share urban public spaces, without any discrimination and restrictions through multiple options and choices. Finally, spatial quality refers the effectiveness and capacity of urban spaces to function the communities properly in built environment with higher performance efficiency of space usage that fulfils the human needs together with spatial justices through effective planning processes like un-indiscriminate application of density in the built-environment. Therefore, spatial quality of urban space is the extent to which that space satisfies the expectations of a community. In relation to the spatial quality of urban space, *density* of built environment has direct or indirect influences on it. The following *table 2.1* depicts that different urban planning practitioners have been defined spatial quality in various aspects

Table 2.1: Theories of defining spatial quality of urban spaces (different scholars)

No	Name of the Scholars who define spatial quality of urban spaces	Spatial justice	Quality planning	Place quality	Good city form	Good design	Defensible space	Fulfillments of human needs	Effective planning process & Planning Performance	Urban quality	Inclusive design	Degree of self-sufficiency
1	Soja, 2010	Equity, fairness, access										
2	Creedy et al., 2007		stewardship, resilience, diversity									
3	Healey, 2004			Sense of place, fit								
4	K. Lynch, 1984				Vitality, fit, access, control							
5	Sternberg, 2000					Integrated, access, fit						
6	Friedman, 2004								Quality planning			
7	Berke et al, 2004								Quality planning			
8	Trip, 2007									Quality amenities, accessibility		
9	O. Newman, 1972						CPTED, many eyes on street					
10	Southworth, 2003										Livable city	
11	Moudaert, 2009							Adequate amenities, accessibility				
12	Lang, 1990										Livable city	
13	Carmona, 2004		stewardship, resilience, diversity									Accessibility, diversity, flexibility
14	Oswald et al., 2003		stewardship, resilience, diversity									Accessibility, diversity, flexibility

Source: Authors Compilation from different references, 2014

2.2.3. Spatial quality and Usability of Space in Built Environment

In cities located within hot dry and semi-humid tropical climates like Addis Ababa, the quality of dwelling environments is mediated by the way houses have been organised to facilitate maximum cross-ventilation and comfortable use of indoor and outdoor spaces. The presence of shades from plants as protective mechanisms from extreme weather conditions is of vital importance to improve the microclimate and enhance comfort living within and outside the dwellings. Kyhn notes that:

For people in the tropics, it is normal to eat, cook, work and play outdoors and to seek shelter of a house only when the need for privacy demands it, and thus the treatment of the adjoining ground should be seen as an extension of the homes. Outdoor life in a warm humid climate is only pleasant if there is a breeze, shade and protection from rain. (Kyhn, 1984:54)

Kyhn argues further that to maximise the breeze there should be no enclosure wall at all. While perforated fencing or screens that obstruct vision may be used, fencing that obstruct air movement should not. Freely grouped houses should be encouraged to make them 'wind transparent'. With regard to the significance of shade trees, Kyhn notes that:

Roof overhangs, verandahs, partios and covered passages are welcome but *the best is a shade from a tree. Shade tree filters the sunlight, reduce air temperatures by evaporation, protect smaller plants and on the ground reduce glare from bright overcast skies.* (Kyhn, 1984:54)

Similar conclusions have been made by Nnkya when she characterises design for comfort in the coastal tropical climate:
A combination of high temperatures and high humidity causes permanent discomfort. However, the monsoon and local winds produce a cooling effect to some extent. Maximisation of cross ventilation indoors and outdoors is thus important. As much as wind or air movement as possible should be directed through indoor spaces and outdoor activities located to utilise breezes. (Nnkya, 1984:79)

With regard to in-door comfort, Kyhn observes that comfort is largely dependent on the control of air movement and radiant heat, the prevention of solar radiation from reaching the building's occupants directly through doors or windows or indirectly by heating the structures. These requirements point to the need for light well insulated construction of walls and roofs, reflective surfaces, correct shading and design for good breeze penetration.

It is imperative from the foregoing that spatial qualities both indoor and outdoor are primarily dictated by the degree of exposure to cross ventilation and shade from trees to buildings and outdoor spaces. However, one of the key parameters in achieving this has been how buildings have been organised and the internal design of buildings themselves. Too compactly laid down houses with double banked house types have poor qualities in terms of facilitating cross ventilation. This study endeavours to analyze spatial qualities in relation to variables of cross ventilation, views and the relative quality of in-door outdoor relationships, degree of exposure of buildings, sun lighting and generally environmental qualities related to issues of amenity and infrastructure components. As regards spaces, Amos Rapoport discusses spatial quality and relates it with space use. He argues that owing to the multi-faceted meaning of space as explicitly discussed above, any attempt to evaluate spatial quality will also have variable meanings.

However, he contends that spatial quality becomes a meaningful concept when it is related to space use.

...space as perceived is variable in many ways. It seems intuitively likely that the definition and evaluation of spatial quality would equally be variable - or more so since this involves values and ideas of the good life held by different groups. Good space in good environment is a function of a given context. For a member of a culture which evaluates space primarily as religious space, for example, criteria which we may value highly may not be relevant. Spatial quality becomes a meaningful concept only when related to definitions of space use. (Rapoport, 1970:51)

Further, Rapoport argues that the first problem in characterising space and therefore deciding whether such a space is good or bad is the difficulty of defining it or rather deciding about the kind of space we are dealing with (Rapoport, *ibid.*). But the question to ask is: What are these requirements to characterize effective utilisation and analyse quality of spaces?

Related to this argument Gehl (1987) on the other hand enumerates quality demands for *effective space utilisation* with regard to space use activities such as **walking, standing and sitting** activities of *comfort* and *protection* as follows:

Quality requirements for walking: These include quality of pavement materials and street surface conditions. In situations where the degree of crowding can be determined freely, the upper limit for an acceptable density in streets and sidewalks with the two way pedestrian traffic appears to be around 10 to 15 pedestrians per minute per metre. That people tend to take direct and shortest routes - diagonal routes are more preferred. Usually pedestrians tend to walk at the edge of open spaces to enable them experience the street facades and the open spaces (Gehl, 1987:136-147).

Quality requirements for standing: Standing activity should be linked with the concept of staying. Standing reveals stationary activities in a public place. People will prefer to stay for residences at niches in the facades, recessed entrances, porches, verandas and planting in the front yards. Availability of standing supports such as bollards facilitates longer stays (*ibid.*).

Quality requirements for sitting: The type of seats is a determinant factor in the choice of a place to sit. Benches or chairs sometime referred to as primary seating's are essential requirements for being able to sit. A seat ought to be easy to sit on and comfortable to sit for a long time. Secondary seating in the form of stairways, pedestals, steps, low walls, boxes and so on are needed when demand for seating is particularly great.

The question of protection: Protection in the uses *of spaces* can be viewed from two perspectives, namely, protection against crime and other vices and protection against unpleasant or extreme weather conditions. It has been argued that if there are many people on the street there is considerable mutual protection (Newman, 1973:78-80; Gehl, 1987). The feeling of uncertainty and risk underlies the heavily traffic jammed streets and low-density low-rise settlements. Unpleasant weather varies from area to area and country to country but more important is the protection from negative climatic conditions such as excessive solar radiation, heat and rainfall.

Spatial quality as applied in this study refers to the basic pre-requisites for effective use of spaces. It refers to the necessary pre-conditions to evaluate utilizations of outdoor spaces and liveability qualities within identified and selected housing settlements. Although these variables as developed by Gehl draws their relevance to many European cities as a reaction to *'functionalism'* and increasing car traffic that dominated city planning and development in the 1960s and 1970s, their relevance in this study lies in the pre-requisites criteria to assess quality of spaces. The argument here is that irrespective of the context, there are basic requirements within spaces that facilitate effective utilisation of such spaces. If such requirements are missing, then utilisation of spaces becomes poor or un-conducive. It is from this understanding author extrapolate the applicability of these pre-requisites within Addis Ababa city context. Generally, the above variables would be affected by the high ground coverage in housing structures, which might narrow the possibility of communal private outdoor spaces, public open and green spaces where people can walk, stand, sit and protect themselves from incidence of crime and so on. Therefore, density is very essential planning aspect that should be used wisely so as to bring about sound, decent and habitable built environment by installing spatial quality variables. And hence, the effects of those above explanatory variables determine the overall situation of spatial quality as response variable in the housing settlements.

Since the study is largely built-environment with respect to *density and spatial quality of housing settlements–* in some extent behavior research, the major research instrument employed in this study was observations, questionnaire survey and measurements as well. It has been used for sampled questionnaire survey. Such kinds of studies have long history. Typical studies in the past have always used behavior mapping as a way of understanding the interaction between people and urban space and its quality (Bechtel et al. 1987). Such an approach is premised on the fact that there is less recorded on how people use urban space and the kinds of dimensions and details that support different uses within such settings (Goli˜cnik and Thompson, 2010). In addition to observed behavior of urban space users and their perception on spatial quality as well as the recording of size and typology of amenities and facilities, a questionnaire was administers to a sample of random observed residents of urban space users in selected case study neighborhoods. This was done to comprehensively capture active interaction between people and urban space to determine spatial quality in residential neighborhoods.

Therefore, the relationship between identified predictory variables and spatial quality as response variable was done and could be represented by binary logistic regression model as shown below in equation (1). The inhabitants express their feeling of spatial quality by saying like or dislike, satisfy or dissatisfy good or bad and hence it has dichotomous characteristics of responses. Accordingly, the *binary logistic regression model* was applied to determine factors that explained why some neighborhood residents define their surrounding urban spaces; spatial quality being good (1) and bad (0). When dealing with a dichotomous dependent variable - the main interest is to assess the probability that one or the other characteristic is present (Peng and So, 2002; Peng et al, 2002). The logistic regression model answers the question what determines the probability that the answer is *good or bad (satisfied/like or dissatisfied/dislike)*. The special features of the model guarantees that probabilities estimated from the logistic model will always lie within the logical bounds of 0 and 1. In other words the probability that an inhabitants picked at random is defines **spatial quality** of housing settlement as **spatially good or bad quality** is not a continuous variable but a discrete one. The logistic regression model can be expressed mathematically as follows (*equation-1*);

Where, Θ = is the dependent variable (i. e. probability that an urban citizen chosen at random is agrees that the Surrounding **urban spaces spatial quality as being 'satisfied=1' or 'dissatisfied=0'**.

$$\theta = \frac{e^{(\alpha+\beta_1 x_1+\beta_2 x_2+...+\beta_i x_i)}}{1+e^{(\alpha+\beta_1 x_1+\beta_2 x_2+...+\beta_i x_i)}}$$

Where α = the constant of the equation

β = the coefficient of the predictor variables.

x = are the explanatory variables and log is the natural logarithm

The aggregate expression of the model can be summarized as above.

$$\text{logit}[\theta(x)] = \log\left[\frac{\theta(x)}{1-\theta(x)}\right] = \alpha + \beta_1 x_1 + \beta_2 x_2 + ... + \beta_i x_i$$

Parameters in logistic regression model would be estimated using the maximum likelihood method. The enumerated regression coefficients represent the change in the logit of the probability from a unit change in the associated predictor, assuming other factors are constant (Gujrati, 2003). There are two main uses of logistic regression including: The first is the prediction of group membership. Since logistic regression calculates the probability of success over the probability of failure, the results of the analysis are in the form of an odds ratio. Logistic regression also provides knowledge of the relationships and strengths among the variables.

2.3. Density and its Implications on Spatial Quality

It is strongly believed that density as an urban characteristic, which affects spatial quality of urban spaces in built environment as this study is focusing on population and built-up density/BAR, FAR/. It is a term used in urban planning and urban design to refer to a quantitative measure of the number of units on a particular area of land, or the number of people inhabiting a given urbanized area. It can also be defined as dwellings per unit area or hectare or square metre as *gross/net dwelling density*. The theoretical background presented by (James, 1967:552), who defines *Net residential Density* as "*the quantity, the average intensity of which is measured by net residential density, is either the population, or the accommodation (in terms of houses, habitable-usable floor area spaces) as a proportion of the housing area, which contains, not a mixture of uses and activities but almost exclusively a single use being housing. That means that it includes spaces covered by dwellings and their gardens, any incidental open space and half the width of*

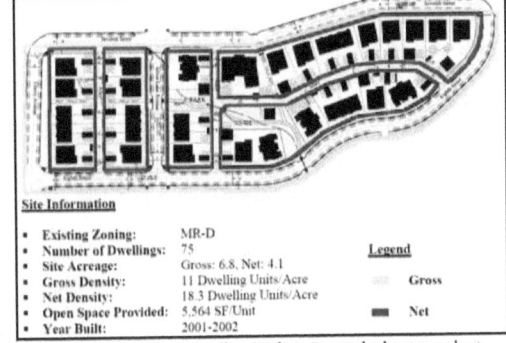

Figure 2.8 : Example of Dwelling Medium Density both gross and net, City Boulder, USA, 2010)

surrounding roads, but excluding local shops, primary schools, and most open space, and all other types of development".

Table 2.2: Dwelling Density and Built-up Density (BAR and FAR) in different Cities at Neighborhood Scale

Name of the City	Density Components at neighborhood scale				
	Residential Density DU/hectare	Size of DU in Persons/DU	Population per hectare	BAR	FAR
Boulder, US	45 to 54 du	3persons/du	135 - 162	40 to 50%	>3.0
Addis Ababa, Ethiopia	84 to 125du	5persons/du	420 - 625	50 to 65%	>2.5
Adelaide, Australia	94 to 120du	3persons/du	282 - 360	30 to 50%	>2.0
Singapore, Singapore	320 to 500du	4pesrons/du	1280 - 2000	30 to 40%	>4.0
Kigali, Rwanda	60 to 100du	5persons/du	300 - 500	40 to 60%	>1.5

Source: Authors Compilation, 2014

The previous sections discussed the use of different density concepts in urban analysis and city building. As Churchman (1999) describes in disentangling the Concept of Density, *there is not one accepted measure of density in or shared by*, different countries (*see table 2.2*). In general, density measures vary according to the manner in which numerators and denominators are defined. Some countries define density using the number of people per given area (population density), while others define it using the number of dwelling units or the building mass per given area (Floor Space Index). It is important to realize that one can translate one density measure into another by making assumptions or applying known statistics, such as dwelling size and the occupancy rate. A purely built-up density, such as FAR, can be translated into a more socially relevant form of density, such as dwelling and population density.

On the other hand, built-up density refers to floor area ratio/FAR/ and base area ratio/BAR/) and is determined by the space between buildings, building width, building configuration and building height. *Gross residential density* refers to the number of dwelling units divided by the total site area, while *net residential density* as explained above refers to the number of dwelling units divided by the area of the site taken up by residential use only *(Fig 2.8)*. *Density is defined not only as intensity (FSI), but as a combination of intensity, compactness (GSI), height (L), and pressure on non-built space (OPR); it can be used to differentiate between urban forms in a more efficient way. (M.Berghauser & P.Haupt, 2010:59)*. On the other hand, the *table 3.2* reveals that how built up and dwelling densities at **neighborhood scale** are varying from city to City.

Therefore, density is considered an important factor in understanding how **cities function**. It is commonly asserted that higher density cities are more sustainable than low density cities. However, the link between urban density and aspects of sustainability remains a contested area of planning theory (Newman, 1999). Many *Planning* experts on sustainable urbanism, including prominent urban designer Jan Gehl, argue that low-density/dispersed cities are unsustainable as they are automobile dependent. At a broader level though, there is evidence to indicate a strong negative correlation between the total energy consumption of a city and its overall urban density, i.e. the lower the density, the more energy consumed (Newman, 1999). This reveals that the efficiency of urban spaces in high density built environment shows greater increase in spatial quality than the low density built environments.

Density can also be expressed in terms of built-up density. It is important to analyze the built-up density, regarding the efficiency of land uses; the cost effectiveness of infrastructure has a direct relationship to the intensity of building density. According to Acioly and Davidson, *"The size of* plot, the amount of plot which can be built up and the height of the building give the dimensions of the most visible aspect of density: the amount of space which is built" (Acioly and Davidson, 1996:7).

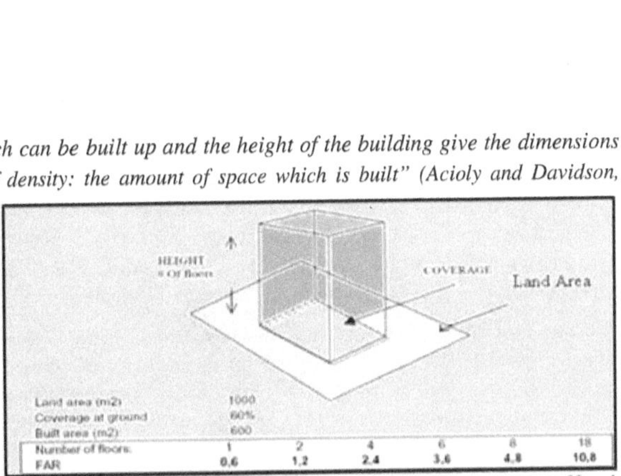

Computations of ground coverage ratio/BAR/ and Total Floor area ratio/FAR/ of the neighborhoods have been done using the following equation:
BAR=BA/PA,
FAR=FA/PA and
BH=FAR/BAR, where BA is ground coverage area, FA is Total Floor area, PA is plot/site area and BH is Building Height, BAR is built-up area ratio/Ground Coverage ratio/and FAR is total floor area ratio. Spatial quality as dependent variable based on the composition of density and some other basic predictory variables including outdoor spaces, accessibility and circulation, parking spaces, green and open spaces, safety and security, ventilation and daylight access.

Figure 2.9: Conceptal Model of FAR values and percentage of land Coverage.

Built-up density includes land coverage/BAR/ and floor-area ratio (FAR). Rådberg(1996) whose approach has been to develop a general framework or methodology using typo-morphological analysis as opposed to functional typology analysis. In the procedure of actual calculations of built-up density (BAR/FAR), actual site and architectural plans have been used. For all sites, coverage of the area are measured and called "site coverage". These are measurements for the "site" which is the actual area of the plot in which a housing complex is built on, and in all cases included the addition of half of the width of the surrounding streets. In some cases the presence of

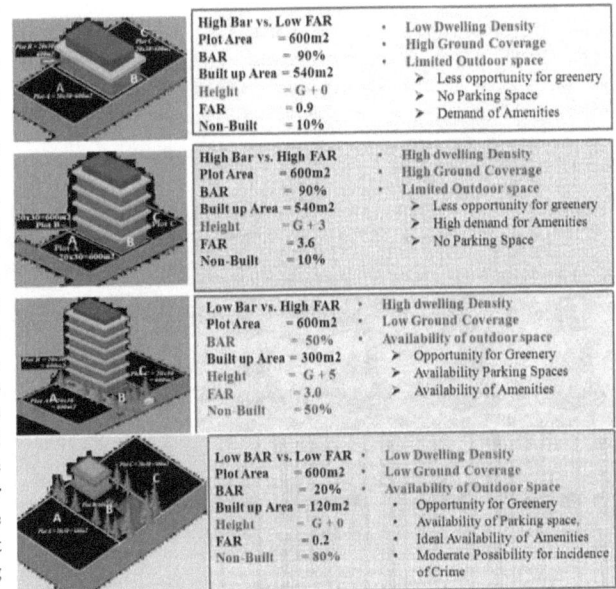

Figure 2.10 : Built-up Densities in Different situations and its implications on spatial Quality
Source: authors sketches and compilation, 2014

public facilities such as small day care centres, or large areas with undetermined uses attached to them have been included, if these have been considered as variables that constitute as being a part of the type under analysis. The calculation comprises: *floor area ratio/FAR/:* ratio of total built residential area to area of land, *building height*: number of stories and *coverage of the site/BAR/*: built up area divided by the size of the plot. The *figure 2.9, 2.10 and table 2.2* reveals that how density varies from city to city which calls for scientific enquiry.

However, a naked eye observation on current urban morphologies in many developing country cities like Addis Ababa, Ethiopia suffices to retrace the minimal role of planning intervention in `urban space consumption` over time as explicitly addressed above. This predominance of haphazard development poses a substantial need for re-planning. The high building coverage's of land contributes for the absence of open and green spaces, circulations, possibility to use outdoor spaces. Therefore, planners and designers should have to pay due attention to develop urban built-form & morphology with thresholds of density as prescriptions and norms to control quality urban spaces in built environment. As a result, planning controls were developed that prescribed *maximum* allowable densities (Churchman, 1999). Many municipalities and design experts have been strived to determine the minimum and maximum dwelling and built-up density as 'norms and prescriptions' to plan and design quality built environment in building spatial quality elements in the built environments

It is also worth mentioning that high building coverage against high and low floor area ratio is the basic characteristics of developing countries cities that leads to less possibility of outdoor spaces, less access to mobility and circulation spaces, creates possibilities for incidence of crime, less opportunity to green public urban spaces in the city. Therefore, analysing the different patterns of the built-forms is quite essential to design and plan quality urban spaces and their influence on the spatial quality of urban spaces. The patterns of different built forms have direct or indirect impacts on the spatial quality of urban spaces. These patterns comprise high BAR vs. Low FAR, High BAR vs. High FAR, Low BAR vs. High FAR and Low BAR vs. Low FAR *(see figure 2.9 and 2.10)*.

The patterns of built form at unit level have direct impacts being reflected on the neighbourhood or block level. For instance, **High BAR vs. high FAR** illustrates the effect on spatial quality of

Table 3.3: Correlation between Density, Patterns of Built Form, Spatial Quality indicators and their implications

No	Case Study area & its Patterns of Built-Form	Spatial quality Indicators						Inferences
		Outdoor Space, green & Open Space	Daylight and Ventilation	Demand for Amenities	Population Density	Mobility & Accessibility	Incidence of Crime	
1	Case Area: Wube-Bereha (High Bar vs. Low FAR)	Limited outdoor space, less opportunity for greenery, limited parking lots	Less access to daylight & Ventilation	High demand for Amenities	High density	Limited circulation and mobility spaces	Less visibility and high incidence of crime	Poor spatial Quality, high population density, Overcrowdness
2	Case Area: Gerji-Suns. (High BAR vs. High FAR)	Limited outdoor space, limited greenery & parking spaces	Less access to daylight & ventilation	High demand for amenities	High density	Limited circulation and mobility spaces	Less visibility and high incidence of crime	Poor Spatial Quality, high Population density,
3	Case Area: Lideta-Fird. (Low BAR vs. High FAR)	Better outdoor space, greenery & parking spaces	better access to daylight & ventilation	Availability of Amenities	High density	Better circulation and mobility spaces	Better visibility and less incidence of crime	Better spatial Quality, high Population density
4	Case Area: Yeka-Ayat (Low BAR vs. Low FAR)	Better outdoor space, much greenery spaces, parking spaces	better access to daylight & ventilation	Availability of Amenities	Low density	Availability of circulation & Mobility spaces	Less visibility and high incidence of crime	Ideal Spatial Quality, low Population density

Source: Authors compilation from review, 2014

Note:

High BAR---High Ground Coverage Ratio/Built-up area Ratio,   Low BAR--Low Ground Coverage ratio

High FAR---High Floor area Ratio,   Low FAR---Low Floor area ratio urban spaces on the figure below. It is strongly believed that built form at plot of plot level has influence on the neighbourhood scale, *figure 2.8 and 2.10 above*, it has been illustrated a neighbourhood with high density -low plot area ratio or *percentage of ground coverage* and high density-high plot area ratio.

In planning practice, plot area ratio/PAR=BAR/ is extensively adopted as a standard indicator for the regulation of land-use zoning and development control. Different plot area ratios for different types of land uses are often specified in urban master plans as a provision of mixed land use. Furthermore, maximum plot ratio is often controlled in the master plan in order to govern the extent of build-up and to prevent overdevelopment. In building design, plot area ratio is widely used in design briefing and development budgeting as it reflects the amount of floor area to be built and, hence, can be used to estimate the quantity of resources required for construction; consequently, it can forecast the financial balance of investment and returns. Building or Site coverage represents the ratio of the building footprint area to its site area. Therefore, *site coverage/ground coverage/BAR/* is a measure of the proportion of the site area covered by the building. Similar to plot area ratio, site coverage of individual developments is often controlled in urban master planning in order to prevent over-build & to preserve areas for greenery and landscaping that is being reflected in the neighbourhood or settlement scale.

The open space ratio/OPR/, which is the inverse measure of site coverage, indicates the amount of open space available on the development site. However, the term is sometimes also expressed as area of open space per person and this measure is used by the planning authority to safeguard a reasonable provision of outdoor space for the population. Apart from plot ratio and site coverage, other density measures, such as residential densities at neighbourhood or quarter scale, can also be expressed in terms of building density.

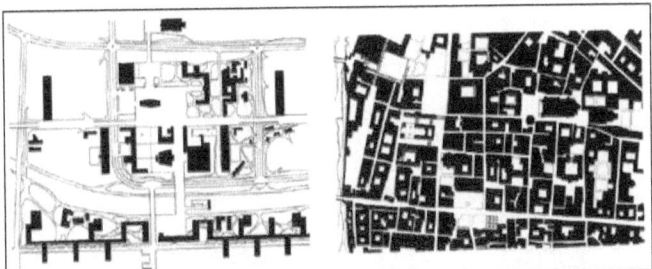

Figure 2.11: Figure-ground analysis from 'Collage City' by Rowe and Koetter (1978).

Measurement of residential density with respect to number of dwellings per land area is an important indicator in the making of planning policy. On the other hand, the concept of coverage was frequently used throughout the 20th century to express the relationship between built and non-built land. *C. Rowe & Koetter (1978)* used the *figure-ground analysis* to visually represent coverage as the distribution of (built) mass and (un-built) open space as illustrated at neighbourhood scale. They used this representation to decode two opposite doctrines at the core of modern and traditional planning: the first an accumulation of solids in an endless floating void, the other dominated by mass and cut through by voids *(see figure 2.11)*.

Generally, patterns of built-forms, 'built-up density(BAR/FAR) at block and plot level, and *'residential density'* at neighbourhood scale in the built environments are the determinant factors, which would affect *'spatial quality'* of housing settlements in terms of access to amenities, outdoor spaces, green and open spaces, ventilation and daylight access, safety and security, Accessibility and mobility. The failure and success of spatial quality may depend on the level of built and non-built spaces in relation to built-up density.

2.3.2. Non-Built Space and Built Total Floor area Ratio

Hoenig was the first to systematically study the density and spaciousness of the urban environment (Rådberg 1988: 68-70). In the article, Hoenig (1928) introduced the concept of spaciousness, defined as the relationship between **open space and total floor area**, as a measurement of the quality of an urban plan. Spaciousness is equivalent to the Open Space Ratio mentioned in the New York City's Zoning Resolution (New York Department of City Planning 1990). OPR was used as an instrument to stipulate that a development must provide a certain amount of open space on a zoning lot in specified districts. It can be viewed as an expression of the trade-offs between the desire to maximize the building bulk (program or FAR) and the public and private demand for adequate green & open space.

At the level of a lot (or building block), as addressed in the literature, it has been proposed a minimum of **one square meter of open space for every square meter of built floor area**. According to the review, when this standard was met, the area could be described as spacious. Built-up areas with less open space were not acceptable and were described as cramped or crowded. In Addis Ababa, the only example that, on the scale of the building block, meets the spaciousness standard proposed by Hoenig was the Lideta-Firdbet. The other Addis Ababa areas analyzed in this research have lower figures. In other neighborhoods, most low-rise samples built from the 1960s to the 1970s fulfill Hoenig's requirements. Other locations do not, but are structurally more spacious than their counterparts in the city. Two samples, Lideta-Firdbet and Yeka-Ayat, with high and low OPR values respectively, consist of large building blocks of 5-8 storeys high and small blocks of 1 to 3 storeys high. The author can thus conclude that OPR alone does not contribute much to the understanding of urban built-form and hence some other variables must be required to apprehend patterns of built –forms including built-up density, building characteristics plot characteristics as determinants of patterns of built-forma and spatial quality. However, it does reveal the character of the areas in terms of pressure on the non-built space. If all of the inhabitants of the dwellings in these houses would go out onto the streets and into the courtyards at the same time, each person would have the same amount of open space at his/her disposal in both samples.

As it has been long argued the concept of open space coverage was frequently used throughout the 20th century to express the relationship between built and non-built land. Colin Rowe (1978) used the figure-ground analysis to visually represent coverage as the distribution of (built) mass and open space as explicitly addressed above. He used this representation to decode two opposite doctrines at the core of modern and traditional planning: the first an accumulation of solids in an endless floating void, the other dominated by mass and cut through by voids, which are very important elements to determine the threshold of open space ratio/OPR/ in the built-environment *(figure 2.10 & 2.11)*.

On the other hand, the relation between street width, or court size, and building height was also a factor in the studies of Walter Gropius in relation to open spaces. He argued that by planning for high density high-rise buildings, one could provide more open space without losing out on the number of dwellings (and population density). Later, Alexander (1977: 114-119), was arguing against the modernist high density high-rise developments, introduced psychological arguments to subject all buildings to height restrictions. Based on evidence from the British Medical Journal and Newman's experience since the early 1970s of carrying out and analysing *'Defensible Space'* projects, Alexander (1977: 115), was dealing with adequate open and green spaces not only for recreation but also for natural surveillances to protect the neighborhoods from crime.

2.4. Safety & Crime Prevention in High Density high-rise Housing Forms

One of the biggest problems in our modern cities is the increasing rate of crime and fear of crime. Government and law enforcement offices, trying to control this phenomenon, have focused most of their efforts in combating it through repressive or police force-related methods. Many years and an enormous amount of money have been spent but the problem is still considered as the main social concern in modern society in relation to improving the quality of life of urban communities. This means that it is time that we consider alternative options for the solution to the problem of criminality in our cities. Instead of combating it, why don't we try preventing it to happen? Oscar Newman, in his book ***"Defensible Space"***, states:

*The crime problems facing urban America will not be answered through increased police force or firepower. We are witnessing a breakdown of the social mechanisms that once kept crime in check and gave directions and support to police activity...Because of the **size and density** of our newly evolving urban megalopolis, we have become more dependent on each other and more vulnerable to aberrant behavior than we have ever been before (Newman, 1973).*

Another important crime analyst, Richard Gardiner (1978), cites as traditional ways to combat crime: police investigation and arrest procedures, criminal justice punishment and threat of punishment, and individual defensive measures. But in recent years, Gardiner affirms, there has been a change in attitude towards crime, and it is more common to find programs and plans focused in prevention of crime such as: *citizens participation in block watch programs, leadership by police in crime prevention programs, community participation in the design and planning of preventive plans, and a closer relationship between citizens and police*. Most of the efforts to combat crime until the 1960's were basically through *repressive methods*, using police and other criminal justice agencies. Just after many years of looking at crime rates rise steadily, governmental agencies began looking for alternative methods for controlling crime, such as opportunity *reduction or situational crime prevention,* which looks for any flaws in *built and social environment* that can contribute to crime. In the early 1980's researchers demonstrated the effectiveness of this method. And more recently, a third method to control crime has been implemented in urban centers: *social crime prevention,* which makes reference to special programs designed to help families and communities in high-risk areas (Bright, 1992: 17).

Newman also states that the majority of crime in cities is merely opportunistic, that means, conditions of the place give the chances for crime to happen. This theory is supported by Wilson and Herrnstein (1985) who explain that any individual is confronted to different degrees of chances to *commit a crime*, and that it is the **density** of such opportunities that make a place *less or more safe than other places*. Conditions that create these opportunities for crime, among others, are: *inadequate outdoors* and *public-spaces lighting, lack of surveillance, spaces hidden from pedestrian or vehicular view*, and *many others, which are* **elements of spatial and physical quality of urban spaces**. Who is responsible for designing and planning these spaces? Mainly architects, urban designers and planners, and landscape architects. Those professionals designing and planning public spaces should look for adequate solutions that not only are aesthetically *pleasing, functional efficient, and financially* viable, but also should design consider **safety and reducing opportunities for crime**.

Important figures in the landscape architectural profession, such as Clarence Stern and Henry Wright, have insisted in the social role of landscape architecture (their housing project in Radburn (1928), *New Jersey, is still today, a very valuable example of a design that has taken into account the social needs of people, besides the functional needs)*. We cannot continue designing in a subjective way when our "client" is the whole society, with special needs, such as **safety and crime prevention**, to be taken into account. Jan Gehl (1987) reminds us that: "*An appreciation of the interaction between the physical design and the social characteristics of housing developments is critical to any security plan*". One basic requirement is the right that *every person has to be safe in his own place of residence*. Our objective, as planners/designers, should be to find a way to provide a solution that is at the same time attractive to investors and developers, and is in accordance to this *social prerogative*.

Many theorists have studied the topic of **crime prevention through planning and design (CPTPD)**. Some of them, such as *Clare Cooper-Marcus, Jane Jacobs, Oscar Newman, and Jan Gehl* among others, have focused their studies on the **physical features that promote safety and discourage crime**. Others, mainly psychologists and sociologists such as *Amos Rapoport, Irwin Altman, and Edward Hall*, have shown how the **spatial relations of spaces and their physical design/plan** can influence human behavior in a positive or negative way. Author thinks both positions are valuable in order to find correct answers in designing and planning **safer spaces** for our cities.

As addressed above increasing surveillance: Surveillance is a well-established factor in criminal activity. Jane Jacobs (1961) suggested that the simple presence of more *"eyes on the street"* would deter crime, and this concept was prominent in Oscar Newman's (1972) classic *Defensible Space* and appeared in Jeffery's (1971) *Crime Prevention through Environmental Design as strongly discussed above*. Since then, many studies have shown that perpetrators avoid areas with greater surveillance and greater likelihood of intervention (e.g., Cromwell, Olson, & Avary, 1991; Poyner & Webb, 1992). And, substantial research has shown that criminals avoid well-used residential areas where their activities might easily be observed (Coleman, 1987; Macdonald & Gifford, 1989).

This on the other hand, a *"defensible space"* is a living residential environment which can be employed by inhabitants for the enhancement of the lives, while providing security for their families, neighbors, and friends, (Newman, 1971:3). The spatial layout of **multi-family dwelling**, from the arrangement of the building grounds of the interior grouping of apartments, achieves defensible space when residents can easily perceive and control all activity taking place within it that can enables residence to take control of their neighborhoods to reduce crime by imposing surveillances through the provision of enough spaces around buildings, open spaces and wide streets (restructuring physical layouts).

There is some evidence to suggest that in inner-city neighborhoods, **vegetation or greeneries** might introduce more eyes on the street by increasing residents' use of neighborhood outdoor spaces.

Figure 2.12: Showing High Density High-rise with low ground coverage, PPS, 2010

A series of studies conducted in inner-city neighborhoods has shown that **treed outdoor spaces** are consistently more well used by youth, adults, and mixed-age groups than are treeless spaces; moreover, **the more trees in a space**, the **greater the number of simultaneous users** (W. C. Sullivan, Kuo, & De Pooter, 2001). Not surprisingly then, a recent study found that children were twice as likely to have adult supervision in green inner-city neighborhood spaces than in similar but barren spaces (A. F. Taylor, Wiley, Kuo, & Sullivan, 1998) *(see figur 2.12)*. Thus, in these settings, higher levels of vegetation not only preserve visibility but may also **increase surveillance** to increase safety and security so as to minimize crime.

Population density has also received considerable attention as it relates to crime. Jane Jacobs (1961) also contradicted the popular wisdom of city planners with her claim that crowded city streets and sidewalks could be effective deterrents to criminal behavior. A number of national studies tested the relationships between density and crime, with differing results, some studies, such as those by Schuessler (1962) or Galle, Gove, and McPherson (1972), found positively correlated relationships between crime and density. *"The additional "eyes on the street" created by the development of East Village in Minneapolis has led to a safer vibrant community"*. Arizona researchers found that when police data are analyzed per unit, apartments actually create less demand for police services than a comparable number of single-family houses. In Tempe, Arizona, a random sample of 1,000 calls for service showed that 35 percent originated from single-family houses and just 21 percent came from apartments. Similarly, a random sample of 600 calls for service in Phoenix, Arizona, found that an apartment unit's demand for police services was less than half of the demand created by a single-family house.

One reason for the misperception that **crime and density** are related could be that crime reports tend to characterize multi-family properties as a single *"house"* and may record every visit to an apartment community as happening at a single house. But a multi-family property with 300 units is more accurately defined as 300 houses. To truly compare crime rates between multi-family properties and single-family houses, the officer would have to count each household in the multi-family community as the equivalent of a separate single-family household. When they do so, many find what the previous studies prove: that crime rates between different housing types are comparable.

Therefore, **Higher-density high-rise** developments can actually help *reduce crime by increasing pedestrian activity* and *fostering a 24-hour community* that puts more *"eyes on the street"* at all times. Many residents say they chose higher-density housing specifically because they felt more *secure there*; they feel s*afer* because there are more people coming and going, making it more difficult for criminals to act without being discovered.

"Attractive, well-designed, and well-maintained higher-density high-rise development with low ground coverage attracts good residents and tenants and fits into existing communities by ensuring safety and security".

Table 3.4: Space and Approaches to Crime Prevention-Situational Approaches

	JANE JACOBS	OSCAR NEWMAN	CPTED	BILL HILLER
Control Of Space/ Territoriality	Clear Demarcation between public and private space	Territoriality-capacity of the physical environment to create perceived zones of territorial influence (including mechanisms symbolizing boundaries and defining a hierarchy of increasingly private zones	Natural access control aimed at reducing opportunities by denying access to the crime target. Territorial Reinforcement Physical design strategies creating or extending a sphere of influence so that users of a property develop a sense of proprietorship	Spaces integrated with other spaces, so that pedestrians are encouraged to see into and move through them.
Surveillance	Need of 'Eyes upon the street' belonging to street's 'natural proprietors' (both residents and users). Enhanced by a diversity of activities and functions that naturally create peopled places.	Surveillance-Capacity of physical design to provide surveillance opportunities for residents and their agents.	Natural Surveillance as the result of the routine use of property.	Surveillance provided by people moving through spaces.
Activity	Sidewalks need 'users on it fairly continuously, both to add the number of effective eyes on the street and to induce people in buildings along the street to watch the sidewalks in sufficient number'.	Rejects the argument that more activity on the street and the presence of commercial uses necessarily reduces street crime.	Argues for reduced through-movement and hence reduced levels of activity.	As feeling safe depends on areas being in continuous occupation and use, areas should be designed to enable this (e.g. by making them better integrated with regard to the movement system.

Source: Carmona et al, 2003 (Situational Approaches)

Opportunity reduction methods within the mainstream urban planning and design literature and involve key themes of activity, surveillance and territorial definition and control as addressed in the above *table 3.4*. Organizing with Jane Jacobs (1961), these ideas were developed through Newman's ideas of *'Defensible Space' and the CPTED approach*. More recently, Bill Hiller has offered a perspective on crime and safety that re-engages with Jacobs. This on the other hand unveils that the spatial quality of urban spaces can make sure the reduction of crime by proper planning of built environment through high density high-rise with low ground coverage measures.

Therefore, planning and design practitioners made sure that there is a correlation between urban density and natural surveillance. An adequate density through proper design and planning is necessary to provide a sufficient number of people to support activities which generate vitality. Furthermore, higher density high-rise generates more flows and movements, which provide natural surveillance on the streets. In low-density low-rise areas, where activities are lacking and flows are weak, safety on the streets and in public spaces cannot rely on natural surveillance, but needs to be provided by other tools such as organized surveillance, semi-organized surveillance (neighborhood watch etc.) or technological means . However, if *density is too high with* **high ground coverage,** other problems may arise such as lack of public spaces and higher risk of conflicts among the inhabitants.

Urban design and planning schemes should provide opportunities for enhancing, among residents and users, the sense of neighborhood and of belonging to the place (squares, shops, landmarks, play grounds, historical symbols, social events etc.). In fact, people take care and develop a sense of respect and protections for places that they feel belong to them. For a good sense of neighborhood it is important however to prevent one group from monopolizing public spaces and excluding other groups. In high-density high-rise areas planning schemes should foresee an adequate provision of public spaces in terms of amount, location, quality and possible use, as the concentration of people without sufficient spaces can increase potential conflicts. Planners should have to be careful not to create open spaces which could turn into no-man's lands. Public spaces should be planned avoiding: Empty or out-of-scale places; large areas with a single use (driving, parking, walking etc) and completely confined spaces with limit. On the other hand, some practitioners and people consider density as Overcrowdness that leads to wrong assumption; however, the fact is as addressed below.

2.4.1. Dilemma of High Density and Overcrowdness along with Spatial Quality

In order to investigate the importance of density within an urban framework of a city and what kind of benefits it generates for its users other than stated in above sections, it is essential to define the concept of density. The idea of urban density has been discussed since the Garden City model in the United Kingdom at the end of the 18th century. In the 1920's this discussion continued influencing urban development. Density is a term often regarded as representation of a physical area and the number of people who inhabit it and the number of building occupied the given unit is of land.

However this term is often associated with overcrowding, which was regarded as an unfit environment. The housing problem was one of the major debates in the beginning of the nineteenth century. Urban planners were persistent to improve the living conditions in densely overcrowded metropolitan areas. For a long period of time density was considered as one of the major ills of the city and in response the urban planners saw low density low-rise as the salvation of the own city. In 1898 Ebenezer Howard proposed the urban Garden City form model, which included only low-density dwellings in its master plan preventing further overcrowding of the city by also restricting the residents to certain viable thresholds. Ebenezer Howard looked at the slums of London, which had too many dwellings per acre and too many people per dwelling unit.

However it is important to clarify that high density and overcrowding are not the same thing but both influence spatial quality of housing settlements.

The significance of these terms is strictly separated. High density means there is a large number of dwellings on a piece of land. A good example would be the city of Amsterdam, Singapore and Hong-Kong, which have very high density due to its plot usage law. Overcrowding means there are too many people in a room or dwelling. For instance fifteen years ago the average space for a person in Shanghai was 6 square meters, which means that a dwelling of 30 square meters could accommodate five people, which fits the description of overcrowding, because too many people are present in a dwelling unit. The overcrowding of dwellings or rooms is still persistent in our world and is a symptom of poverty or discrimination. The Garden city movement did not make the difference between overcrowding and high density.

The Garden City planners put these two terms, overcrowding and density, in the same category. The confusion continues until recently reviewing the script by Sir Raymond Unwin, one of the Garden City planners, titled *"Nothing Gained by Overcrowding"*. The text is promoting the benefits of the Garden City Model, which was the response to the overcrowded city. The text generally presents examples and benefits on how to keep *plot ground coverage or built-up area ratio* /BAR/ at an efficient ratio in order to prevent overcrowding. To say that an x number of dwelling units per acre will prevent overcrowding is absurd, because one thing has nothing to do with the other. The Garden City Movement recognized *overcrowding of dwellings by people and overcrowding of land by buildings as the same*, an unhealthy environment for citizens. Therefore a combination of city and countryside was convincing and attractive according to his model in the modern planning so as to bring about comfort and protection in the built environment.

The author needs to refer to the French architect Le Corbusier and his scheme *"Radiant City"* from 1935, used in order to complete the variable between low-density Garden City and high-density Radiant City. The Radiant City was considered as high density high-rise, because its skyscrapers had a high-density core, meaning that each building had a high coverage of inhabitants within extreme *low ground coverage ratio/BAR/*. However the Radiant City is low-density, because the land usage is in the proportions of five per cent dwelling units and ninety-five percent open lands and transport, ideal plot ground coverage/BAR/. The conception of low-density in both cases is absolutely identical. In regard to the term of low-density Garden City and Radiant City have the same character they just use different ratios when it comes to land usage. *Obviously urban density and overcrowding are different terms and can't be put in the same context.* This idea on the other hand strengthens that the problem *of ill planned density built-environment could be resolved by properly designed density in the city or* the problem of density would supposed to be solved by proper consideration of dwelling and built-up/edificatory density in the built environment in order to install good spatial quality elements in the housing settlements.

CHAPTER 3: ANALYSIS AND DISCUSSION

3.1. Introduction

In this section, emerging issues from the four case studies analysis of the Settlements are compared. The results from these case studies analysis are interpreted and discussed so as to establish whether patterns of the issues discussed can be related to each other for all the cases or can be perceived different scenarios in the built-environments. For the purposes of consistency, the major themes used to analyze the four case study settlements with different *densities, patterns of built-forms* and spatial quality of urban spaces are maintained. These include the present *built-form patterns*, *density* and determinants of patterns of built-forms, the relationship between population density and built-forms, Impact of built up density and built-forms on spatial quality as well as safety and security. Across the themes, density, safety and security, issues of patterns of built-forms and their impacts on spatial quality are discussed.

The cross case analysis also explores from the selected case study locations in light of theories discussed in review and the four case analysis to fulfill the aims and objectives of this study.

3.2. Determinants of Built-Form Patterns and Impacts on Spatial Quality

3.2.1. Building height

With the exception of Lideta-Firdbet and Gerji-Sunshine, the majority of the houses in Wube-Bereha and Yeka-Ayat are single storey houses *(Figure 3.1)*. Yet, within Lideta-Firdbet where redevelopment into multi-storey houses is more pronounced, single storey houses are fewer than multi-storey houses. Out of 26 houses that were studied in Lideta-Firdbet, all 26 were multi-storey detached and semidetached condominium buildings.

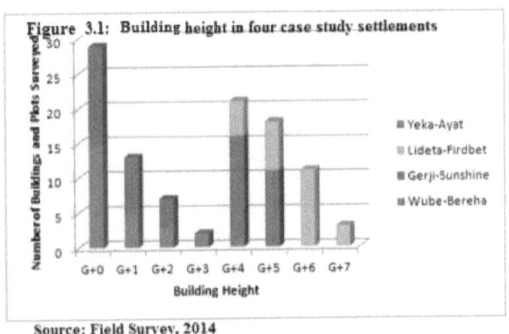

Figure 3.1: Building height in four case study settlements

Source: Field Survey, 2014

The condominium building is rather unique with its 5-8 storeys providing office accommodation to business activities in addition residential uses in the ground floors those buildings laid at the street sides, the apartment owners and a number of other offices rented to private and public institutions.

In Gerji-Sunshine study neighborhood, out of a total of 26 houses, 26 houses were multi-storey, 15 four to five storeys and 11 six storeys. However, Wube-Bereha and Yeka-Ayat are predominantly low-rise housing area. Out of 26 houses in Yeka-Ayat, only 4 were two to three storeys and the remaining houses are single storey. In Wube-Bereha, only three were multi-storey houses as the remaining were single storeys. Variations in building heights for the four case study settlements have been summarized in *figure 8 above*. Hence, Lideta-Firdbet and Gerji-Sunshine are high density high-rise built structures with different patterns of built-forms and had direct impact on spatial qualities. The built-form pattern in Lideta-Firdbet is *low ground coverage and high floor area ratio/high FAR/*, where as built-form pattern in Gerji-Sunshine is *high ground coverage and high floor area ratio/high FAR/* as shown in figure above.

The Redevelopment of high-rise buildings in Lideta-Firdbet has to be linked with the building uses and new demands that arise in the settlement. Since Lideta-Firdbet is part of the city centre, new demands for commercial, office and residential spaces have prompted property developers to buy and redevelop single storey Addis Traditional houses into high rise blocks of condominium and apartments. The fact that former owners of the old houses in Lideta-Firdbet do not have the financial capacity to redevelop their houses, the market forces is replacing these owners through *'redevelopment'* process.

The emerging high-rise buildings in Lideta-Firdbet are largely a result of the redevelopment process and initiatives by government who want to subsidize by responding to these new demands. However, the question that remains is how to balance market forces on land and housing development and ensure acceptable *spatial and environmental qualities*. Perhaps it is easy to argue given marginal role local authorities are playing to co-ordinate housing in Lideta-Firdbet and have got good result that good spatial quality has been installed in the settlement. The reverse is true for multi-story development in Gerji-Sunshine settlement with high building coverage which leads to lose of spaces for public amenities, outdoor spaces, and green and open spaces, ventilation and access to daylight. This is however a thorny issue that ought to be resolved before extreme negative consequences is going to be experienced in Gerji-Sunshine housing settlement. The analysis result indicates that High-rise building structures facilitates possibilities to protect the neighborhood by creating more pedestrian movements as *'many eyes on the street'* to install natural surveillances so as to minimize incidence of crime and fear of crime as well. So, it could be considered as determinants of built-form patterns and spatial quality.

3.2.2. Plot characteristics

A plot characteristic is one of the determinants which influence the patterns of built-forms in the built environments in various aspects. Plot configuration, that is the size and the shape, influence the built-form of houses and its impacts on the spatial quality. From the four main variables, plot exposure is being discussed to characterize plot configuration from case study settlements. These variables are plot size, plot ratio or plot dimension, plot exposure and plot boundary definition *(see figure 3.2)*, however this article focused on the plot exposure and analysed its impacts on patterns of built-forms and spatial quality.

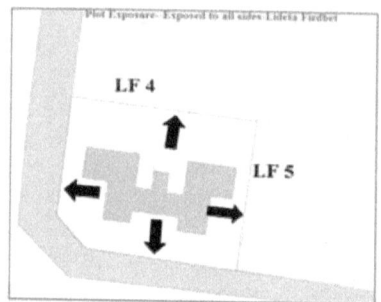

Fig. 3.2: Plot Exposures in All directions (Lideta-Firdbet) facilitates for better ventilation

Since, the underlying assumption in the analysis of exposure is that the more the number of exposures, the more likely that comfort living characteristics are ensured. Empirical observations from the four cases show that in both formal and informal settlements, many plots have no exposure or have limited exposure to only one side. In Wube-Bereha, for example, out of 26 plots, 11 plots have no exposures and 10 plots had only 1 exposure, 3 had two exposures and only 2 had three exposures as Wube Bereha was characterised by high density low-rise with *high ground coverage patterns of built-form*. The basic reason for the absence of exposures in this neighborhood was due to occupation of plots by housing structures dominantly. In Gerji-Sunshine is another neighborhood characterised by *high density high-rise with high ground coverage patterns of built-form*, out of a total 26 plots, 20 plots had two exposure, 6 plots had three exposures.

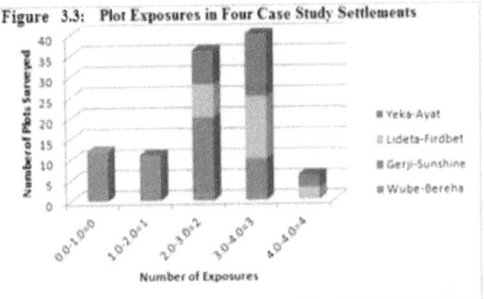

Despite the compact layout of Gerji-sunshine, at least each plot has two exposures. The number of exposures in Gerji-Sunshine is due to the fact that at least each house is facing an access street a bit better than Wube-Bereha's existing situation. Two views can be discussed with regard to plot exposure in Wube-Bereha. First, since Wube-Bereha is too consolidated with compact layout of buildings, the limited exposures for more than half of the plots has made living in this settlement uncomfortable due to the lack of cross ventilation and to some extent inadequate light into the rooms. Second, often views have been blocked due to congested buildings. The overall situation is that amenity within Wube-Bereha settlement is rather poor when plot exposure is compared across cases.

It is evident that the basic characteristics of housing forms in Addis Ababa and most African cities are well known with the limitation of plot or building exposure, which can offer access to ventilation & daylight within and around the building. Two of the case studies have limitations of exposure; especially Wube-Bereha is mainly with the problem of Exposure *(see figure 3.3)*. On the other hand, Lideta-Firdbet and Yeka Ayat have sufficient exposures allowing ventilation and daylight access as good indicator of spatial quality as shown in the *(figure 3.2 and 3.3)*. And hence, Plot or building exposure is one of the important characteristic that has direct impact on the patterns of built form and spatial quality of housing settlements.

This situation is mainly attributed to extensive development of houses and with buildings covering almost the entire plots that have barely left space for alleyways or streets and amenities. Since development in this settlement takes place informally, individual tendencies towards high plot coverage does not take into consideration the need for plot exposure. Apart from narrow footpaths that cannot provide adequate space for exposure, which is quite indispensible for cross ventilation and access to daylight adequately , the remaining unbuilt spaces within the settlement are too few to guarantee adequate cross ventilation to many of the blocked houses. The lack of streets in the inner parts of the settlements further limits the number of exposure in this settlement. For example, more than half of all sampled houses in Wube-Bereha, that is 22 out of 26 houses, do not have vehicular accessibility. Coupled with congested houses, liveability qualities within Wube-Bereha are relatively poor when compared to other case study areas. On the other this situation also lays base for increasing rate of the incidence of crime in the neighborhood as well as highly inaccessible to protect fire accidents if it happens.

3.2.2. Density characteristics and its impacts on built-forms

3.2.2.1. Ground Coverage and Floor Area Ratio

When plot or ground coverage was calculated across the cases, Wube-Bereha reveals a relatively higher coverage ranging from 85% to about 110% *(see figure 3.4 and 3.7)*. Plot coverage for Yeka-Ayat is very low (21%-35%). Plot coverage in Gerji-Sunshine spreads across the range that is between 75% and 90%. Low plot coverage is also notable in Lideta-Firdbet, which is within the standards of the city Administration *(Figure 3.4)*. While Yeka-Ayat was designed as a low-density area, whereas higher plot coverage in Wube-Bereha and Gerji-Sunshine were a result of market forces to maximise profit from rental accommodation by individual developers.

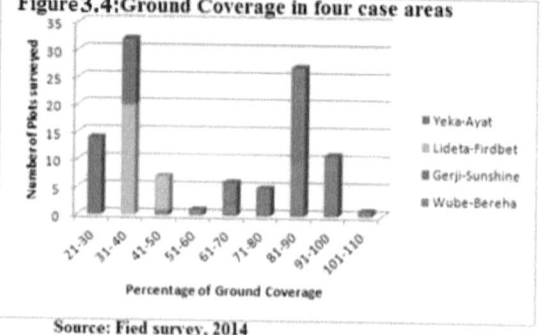

Both Floor area ratio and ground coverage of the plots are the basic characteristics of density, which have direct impact on the patterns of built forms and spatial quality of housing settlements. It is strongly believed that both high ground coverage with low floor area ratio and high ground coverage with high floor area ratio as built-forms would highly threaten the possibilities for the provision of public amenities like recreational spaces, greeneries, open spaces, walking and sitting spaces as well as affect the ventilation and circulation of air within buildings severely. Whereas low ground coverage with high floor area ratio can render the possibilities for the provision of public amenities in the built-environment sufficiently. The figure illustrates the density characteristics in four case study settlements with different patterns of built-forms, which influence the spatial quality positively or negatively in the built-environment.

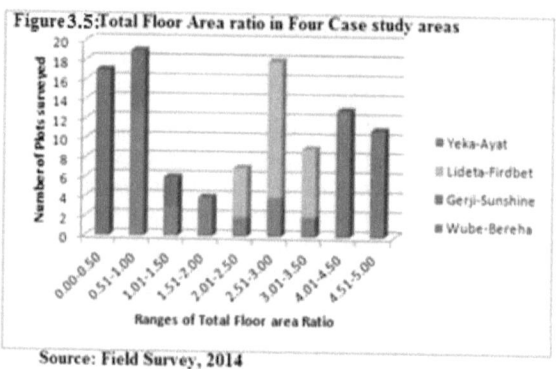

37

The *Figure 3.4, 3.5 & 3.8* also illustrates that *Lideta-Firdbet high density high rise with low ground coverage* has better quality urban spaces than *Gerji-sunshine high density high-rise with high ground coverage* rations/BAR/ because most of the spaces were occupied by building structures. Similarly Wube-Bereha high density low-rise with high ground coverage has poor quality urban spaces than low density low-rise with low ground coverage, however in *Yeka-Ayat low density low-rise with extreme low ground coverage built-form* there were very loose interaction and security problem. When floor area ratios at plot level are considered, the result is generally low ratios ranging between 0.35 and 0.98 at Yeka Ayat, Wube-Bereha has ratios ranging from 0.5 to 1.15 *(see figure 3.7 and 3.9 respectively)*. Both settlements are low-rise housing forms with different density characteristics that Wube-Bereha is high density low-rise characterised by overcrowdness and Yeka-Ayat is low density low-rise with sparse spatial layout.

It is indicative that the general pattern in floor area ratios at plot level across the cases is generally low, medium and high with significant variations. This observation can be attributed to the single storey character of buildings particularly in the Wube-Bereha consolidated informal and formal settlements and relatively larger plot sizes in Yeka-Ayat, even though there are houses with two to three storeys in Yeka-Ayat, actually very in number.

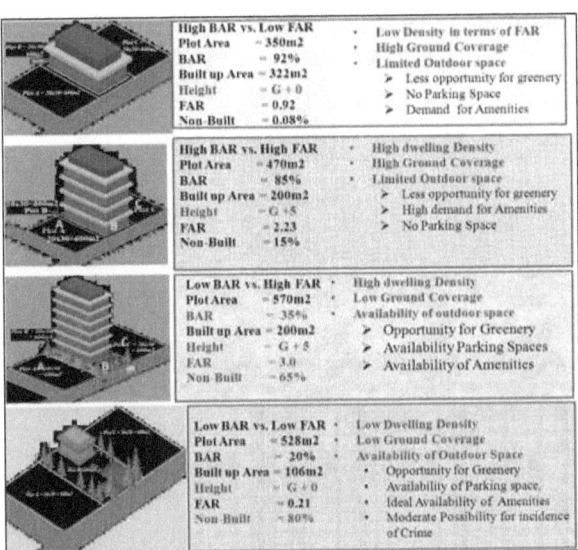

Fig. 3.6: Built-up Density with different patterns of built-forms at plot level and different ground coverage

Therefore, the analysis result unveils that high ground coverage and low floor area ratio have been influenced the patterns of built-forms and spatial quality of urban spaces by limiting possibilities for the provision of adequate public amenities, outdoor spaces, ventilation, green and open spaces, possibilities to combat incidence of crime, accessibility and circulation in the built environment. From the above cross case analysis, Lideta-Firdbet characterised by *high density low-rise* with relatively *low ground coverage* pattern of built-form has been offered better spatial urban spaces than the other three case study neighborhoods in building the above stated spatial quality elements in the built-environment.

3.3. The Effects of Total Floor area Ratio and land coverage at Block & Neighborhood level

Density is expressed as number of houses per hectare, occupancy characteristics, and plot coverage and plot floor area ratio provide an explicit magnitude of intensity of development of spatial quality and their impacts on the patterns of built-forms especially where there are houses with more than one storey as the case is for Gerji-Sunshine. The number of houses per hectare is therefore misleading if parameters of built forms are not defined and vividly identified. For example, there are buildings with more than 6 storeys in Gerji-Sunshine, but when a housing unit per hectare is used to calculate density, and a house with several floors is counted as one, this will be misleading.

Number of Storeys: 1-3
Land Coverage: 85-110%
Floor Area ratio: 0.87- 0.1.15

Row houses in Wube-Bereha area. Note that two units of the row house share the toilet and kitchen facilities at the back of the main building, High Density low rise

Fig. 3.7: High density low-rise with high BAR built-form, crowded and poor spatial quality, ventilation by penetrating roofs. No space for circulation (compiled by Author, 2014)

It is the floor area ratio at a block and neighborhood level that provides the dimension of intensity of development of an urban built-form. To unveil density variation in the case study areas, two variables of land coverage and floor area ratios at block level are examined. Density at block level includes those

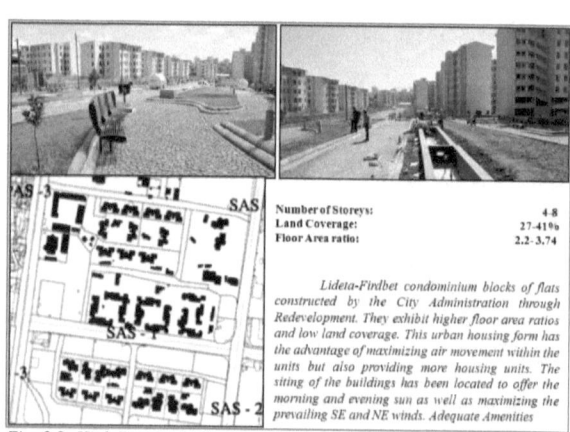

Number of Storeys: 4-8
Land Coverage: 27-41%
Floor Area ratio: 2.2-3.74

Lideta-Findbet condominium blocks of flats constructed by the City Administration through Redevelopment. They exhibit higher floor area ratios and low land coverage. This urban housing form has the advantage of maximizing air movement within the units but also providing more housing units. The siting of the buildings has been located to offer the morning and evening sun as well as maximizing the prevailing SE and NE winds. Adequate Amenities

Fig. 3.8: High Density high-rise with low BAR built-form pattern, comprises better spatial quality elements, morphologically good space layout (by author, 2014)

facilities that are usually part of the daily requirements of urban settlement. Together with developments on plots, it includes half the width of the surrounding roads and services such as shops, incidental open spaces at cluster or block level. On the other hand, the patterns of built form at unit level have direct impacts on the neighbourhood and block level. For instance, **High BAR vs. high FAR and high BAR vs. Low FAR** *(figure 3.7 & 3.10)* illustrate the effect on spatial quality of urban spaces. It is strongly believed that the effect of built form at plot level is being reflected on the neighbourhood scale as illustrated on the sketches.

When blocks are employed as units of analysis, land coverage and floor area ratios in the four case study areas show variations within and between cases as illustrated in Figures above. It is apparent that while Wube-Bereha reveals higher land coverage than the other three cases but also with variations within the case, it has low floor area ratios compared to Gerji-Sunshine and Lideta-Firdbet.

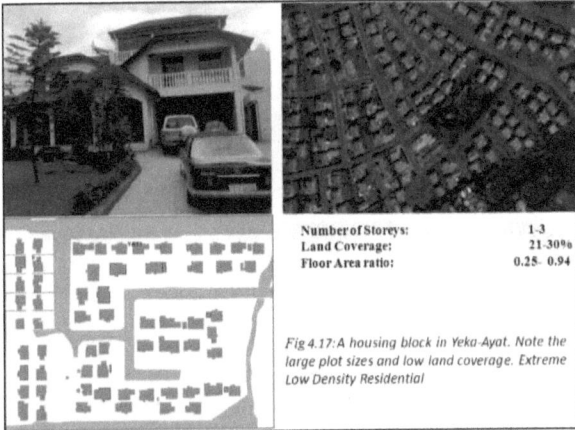

Fig 4.17: A housing block in Yeka-Ayat. Note the large plot sizes and low land coverage. Extreme Low Density Residential

Number of Storeys: 1-3
Land Coverage: 21-30%
Floor Area ratio: 0.25- 0.94

Fig. 3.9: Low Density Low-rise with extreme low BAR, very sparsely settled and residents suffering from incidence of crime, frequent robbery and burglary (by author, 2014)

Variation in coverage is related to the amount of open land that is either yet to be built or presently existing as informal squares or un-built courtyards, and streets within the settlement. Little variations in terms of land coverage and floor area ratio can be noted in Gerji-Sunshine and Yeka-Ayat. Wube-Bereha and Gerji-Sunshine

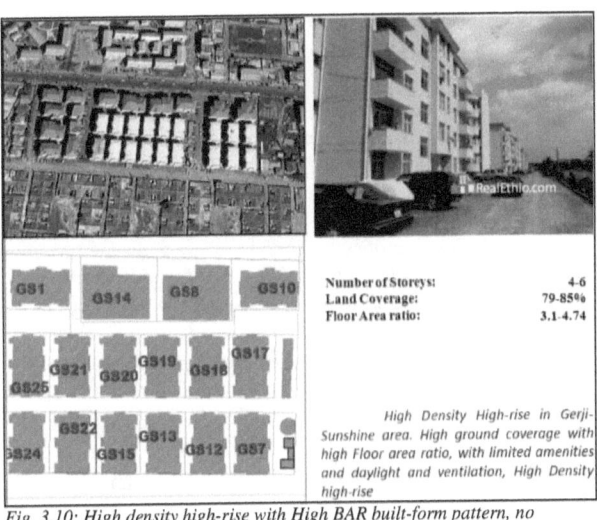

Number of Storeys: 4-6
Land Coverage: 79-85%
Floor Area ratio: 3.1-4.74

High Density High-rise in Gerji-Sunshine area. High ground coverage with high Floor area ratio, with limited amenities and daylight and ventilation, High Density high-rise

Fig. 3.10: High density high-rise with High BAR built-form pattern, no adequate spaces for amenities, parking etc. (Gerji-sunshine), (by author, 2014)

show negligible variation in land coverage. If horizontal extension and land coverage are taken into consideration, then Wube-Bereha*(figure 3.7 and 3.11)* portray rather horizontally densified urban settlement whose negative consequences have been more apparent than in Lideta-Firdbet*(figure 3.8)* and Yeka-Ayat *(figure 3.9)*. When vertical densification and increase in floor area ratio is considered, Lideta-Firdbet and Gerji-Sunshine *(figure 3.10)* prominently depict a vertically densified urban settlement that optimizes land but Gerji-Sunshine with high percentage of ground coverage depicting negative externalities associated with unregulated vertical densification that has been clearly affecting patterns of built-form and spatial quality of housing settlement.

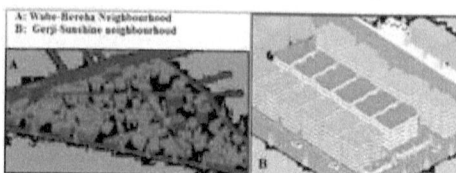

Fig. 3.11: The effect of built-form patterns at plot level is being reflected at neighborhood level, or high BAR at plot level can reflected at neighborhood scale (by author, 2014)

3.4. The Impact of Built-Up densities on Patterns of Built-form

As it has been introduced so far, one of the objectives of this study was to analyze the built-up densities in terms of floor area ratio/FAR/ and percentage of ground coverage by buildings of housing settlements and its impacts on spatial quality. According to Rådberg, the parameters that can be used to measure urban physical density are: residential density, building height and percentage of built up area (Rådberg, 1996:390). The residential density which means FAR and the percentage of land coverage by building are being considered in this study to measure the built-up densities. FAR is the ratio between total floor area by number of floors and the land and plot area. The total floor area means the area of total floors occupied by all the buildings available in the block. The land area includes the total land area covered by block with half of its surrounding roads width. Percentage of ground coverage/BAR/ is the percentage of total land covered by buildings inside the block and the total land area of block with half of its surrounding roads.

The analysis of FAR and percentage of land covered by buildings/BAR/ have been done according to the theory of Rådberg which is about the classification of patterns of built-forms. The FAR values which have been found from the measurements for the settlements show that the settlements containing very low FAR values where as the percentage of ground coverage by houses inside the blocks are very high. On the other hand the Lideta-Firdbet settlement containing FAR values ranging from approximately 2.1 to 3.7 which can be considered as a medium to high where the percentage of land coverage is ranging from 21% to 41% *(see figure 3.12)*. The characteristics, in terms of built-up densities and spatial qualities of the most housing settlements in Addis Ababa city are similar. But there is a variation in terms of Built forms due significant variation in plots. The built-up densities, spatial qualities and built forms of Lideta-Firdbet are different to most of the settlements in Addis Ababa city where the condominium houses are made by concrete materials with adequate amenities, good ventilation, safety and security, adequate communal outdoor spaces, Moderate transport accessibility relatively with other three case study neighborhoods.

On the other hand, the houses in Wube-Bereha are made by earth materials and tin's roof, mud and wattle. It is based on this that the two different settlements were selected as a case. The first case of the informal settlements is Wube-Bereha where FAR value is ranging from 0.81 to 0.1.15 can be treated as a very low dense, in terms of FAR value. Almost all of the houses in this settlement are 1 storey *(see figure 3.1 and 3.12)*. The percentage of land coverage by houses are very high like 85-110%. The second case settlement is Gerji-Sunshine where FAR value is high, for instance maximum 3.5-4.78. In this case ground coverage by buildings is maximum 79%-85% which is also very high for any residential area next to Wube-Bereha. All of the houses are multi-story in this settlement. In both cases the percentage of ground coverage by houses is very high which means that there is a shortage of space inside the block to provide

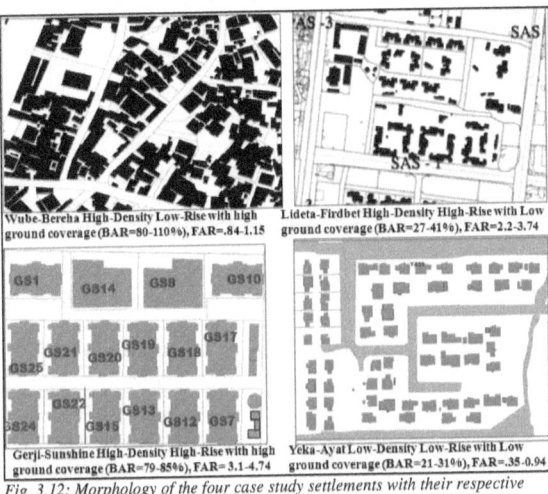

Fig. 3.12: Morphology of the four case study settlements with their respective BAR & FAR, showing each case study areas pattern of built-forms (compiled by Author, 2014)

amenities, private and communal outdoor spaces so as to maintain good spatial qualities and liveability. The FAR value is very low for Wube-Bereha as well; however, the FAR value is high for Gerji-Sunshine, so the FAR value can be increased to increase the efficiency of space, if the ground coverage would be kept optimum. And hence, In order to have good spatial quality, high FAR value should be supported by low BAR/ground coverage in the built-environment.

On the other hand the FAR and percentage of ground coverage by buildings varies from block to block in the settlements. Those depend on the income level of the residents providing the justification for selecting the blocks according to the income levels of the residents. Lideta-Firdbet, the first redeveloped condominium residential area in Addis Ababa city that the FAR value of 2.1-3.74 where the percentage of ground coverage by buildings is approximately 21 to 45%(*figure 3.4 , 3.5 and 3.12)*.. The building height of this block is ranging from 5 to 8 storeys where the FAR value is high and percentage of ground coverage is low. In this settlement the open space is sufficiently provided for the children to play, adults to relax and recreate in the neighborhood. The second formal settlement is Yeka-Ayat high middle and high income inhabitants where the approximate FAR value is 0.35-0.98, which is extremely low and the land coverage is approximately 21-31%. Here the height of building is ranging from 1 to 3 storeys *(See figure 3.4 , 3.5 and 3.12)*. The FAR value is low due to the extreme low land coverage by buildings. In this settlement there is a sufficient private outdoor space in individual compounds and it has no problem of ventilation and access to daylight, however, there is high problem of security as the residents witnessed during interview that there is high rate of incidence of crime and burglary. Even frequent killings have been committed in this neighborhood as indicator of poor spatial quality of urban spaces.

3.5. The Relationship between Population Density & Built-Form Patterns

Built-up density in terms of floor area ratio/FAR/ has a direct relationship with the number of inhabitants residing per unit area or hectare of land. It is crystal clear that FAR value is a ratio between the plot size and the number of times that any one is permitted to cover built-up area as ground coverage is the footprint of that covered area. On the other hand, Floor area ratio can directly determine population density in the built-environment. The higher the number of total floor area ratio, the higher the population number like Lideta and Gerji-Sunshine where high density high rise housing settlements with different scenarios of built-form patterns *respectively*.

As analyzed above, the built-form pattern of Lideta-Firdbet is low ground coverage with high floor area ratio */low BARS vs. high FAR/* where as Gerji-sunshine is high ground coverage with high floor area ratio */high BAR vs. high FAR/* respectively. But in Gerji-Sunshine it evident that due to high ground coverage there were high demand of amenities and services as opposite Lideta-Firdbet. However, one can also perceive *high population density* in the built-form of extreme *low floor area ratio* as different scenarios like Wube-Bereha. This is clear manifestation of high rate of room occupancy in the housing settlement that means many people living in single room or above the recommend minimum 2 persons per room/$22m^2$. This is different from population density of the housing settlement. Population Density is a number of people per unit area or hectare of land, whereas Overcrowdness is the number of people per room or per dwellings exceeding the recommended minimum capacity. The case of Wube-Bereha is typical example of Overcrowdness that the population density is 625 per hectare within extreme low floor area ratio in the built-environment. As analyzed above, the percentage of ground coverage is very high but total floor area ratio is too low, which is clearly showing high dense in terms of ground coverage and low dense in terms of floor area ratio. High ground coverage means the close distance between housing structures that highly threatens the spatial quality by narrowing down the possibilities for outdoor spaces, greenery and open spaces, ventilation and daylight access, visibility, safety and security. The opposite is true for Lideta-Firdbet in terms of spatial quality.

Therefore, built-up density in terms of floor area ratio has been addressed explicitly above as one of the density characteristics that would have determined population density in the built-environment and hence floor area ratio/FAR/ value is directly correlated with population density. According to the analysis result, population density might be high with low floor area ratio lead to Overcrowdness and high population density with high floor area ratio, means much number of people living in multi-storied building structures per hectare. As planner/designer we should have to distinguish Overcrowdness and density in the built environment. Architects and planners should have to prescribe the density thresholds for built environment to maintain spatial quality of housing settlements to create possibilities for adequate outdoor spaces, green and open spaces, ventilation and daylight access, safety and security and the likes as variables of spatial quality elements.

It is also worth mentioning that by keeping population density fixed, one can design built environment with different patterns built forms which might influence spatial quality of housing settlement like High rise with low ground coverage, low-rise with high ground coverage and medium-rise with medium ground coverage categories of dwelling forms in built-environment (*see figure 3.13*).

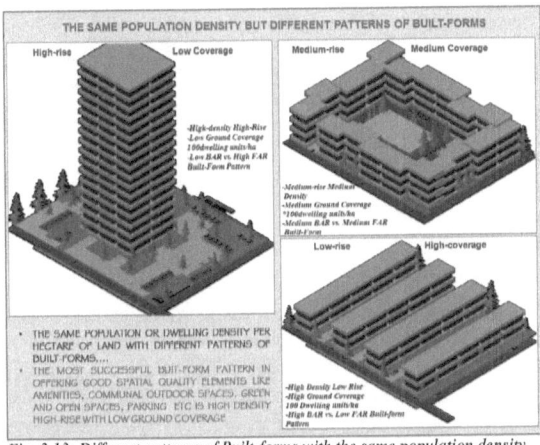

*Fig. 3.13: Different patterns of Built-forms with the same population density, but **high-density high-rise with low BAR** built-form would offer better amenities and spatial quality elements (compiled by author, 2014)*

Therefore, with different scenarios of built-forms, population density can be remaining the same; however the level of spatial quality is different for each built-form pattern. For instance, high density high-rise with low ground converge built-form can offer higher quality urban spaces in the housing settlement than high density low-rise and high-rise with high ground coverage built-form by creating better possibilities for public amenities, green and open spaces, visibility, ventilation, safety and security and likes as illustrated in figure 4. These figures clearly demonstrates that the same *dwelling density* with different built-forms including high density high-rise with low ground coverage, medium density medium rise with medium coverage and high density low-rise with high ground coverage. Each built-form patterns has different characteristics and implications like high density high-rise with low ground coverage has offered possibilities to provide better spatial quality elements in the neighborhood as stated above, whereas in low rise high ground coverage occupied larger spaces by building structure, which has limited possibilities to offer important elements of spatial quality of the neighborhood for the residents at large. This is clear manifestations of how far different *patterns of built-form* with the same *population/dwelling density* are affecting spatial quality of the settlement. Therefore, that is why pattern of built form is one of the determinants, which influence the spatial quality of urban spaces in the city.

3.6. Safety and Security in Higher Density Built Environment

Most people assume that high-density high-rise development generates more traffic safety than low-density development and that regional traffic will get worse with more compact development. In fact, the opposite is true. Although residents of low-density single-family communities tend to have two or more cars per household, residents of high-density apartments and condominiums tend to have only one car per household. And according to the study using data from the case study neighborhoods in relation number of trips per dwelling forms (2013) doubling density decreases the vehicle miles traveled by 38 percent.

The reason is that higher-density high-rise developments make for more walkable neighborhoods and bring together the concentration of population required to support public transportation. The result is that residents in higher-density high-rise housing forms make fewer and shorter auto trips than those living in low-density low-rise housing forms. Condominium and townhouse residents average 5.6 trips per day and apartment dwellers 6.3 car trips per day (*figure 3.13*), compared with the ten trips a day averaged by residents of low-density low-rise communities.

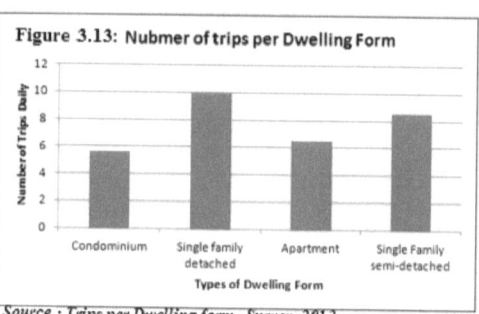

Figure 3.13: Nubmer of trips per Dwelling Form

Source : *Trips per Dwelling form, Survey, 2013,*

A trip is defined as any time a car leaves or returns to a home safely. Increasing density can significantly reduce dependency on cars, but those benefits are even greater when jobs and retail are incorporated with the housing. Such mixed-use neighborhoods make it easier for people to park their car in one place and accomplish several tasks, which not only reduce the number of car trips required but also reduce overall parking needs for the community. Therefore, planned and designed high density high-rise dwelling forms can bring space efficiency in terms of '*safety*', directly the indicator of the spatial quality of urban spaces.

With a typical family now making more car trips for family, personal, social, and recreational reasons than for commuting to work, reducing the number of non-commuting trips takes on greater importance in the battle to reduce traffic congestion and parking problems. A case study in Addis Ababa, Wube-Bereha, found that workers in dense downtown Addis made 80 percent of their mid-day trips by foot while suburban workers made 67 percent of their mid-day trips by car to Ayat-Suburb. Although a suburban

Figure 3:14: High density high-rise with low ground coverage, 2013, Google image, CMC, Ayat, Addis Ababa, Ethiopia

office park would never reach the density levels of a downtown area, planners can still reduce the auto dependency of suburban office workers by using some of the same design techniques. Concentrating density around suburban offices, allowing and encouraging retail and restaurants in and near the offices, and planning for pedestrian and bike access can all reduce the number of lunchtime car trips required by office workers. Higher-density high-rise mixed-used developments also create efficiencies through shared parking *(see figure 3.14)*. For example, office and residential uses require parking at almost exact opposite times. As residents leave for work, office workers return, and vice versa. In addition, structured parking becomes feasible only with higher-density high-rise developments to install *'safety'*. On the other hand, the security of high density high-rise dwelling forms far better than low density housings by creating opportunities for community surveillance through *'many eyes on the street'*. It is also worth mentioning that proper design and planning of built environment to install *safety and security* as entities of spatial quality.

Therefore, density is an important factor to consider in designing safer environments. Nevertheless density, as race, poverty, and other similar qualifications, should be analyzed considering other factors that also play a role in promoting crime activity. It is strongly believed that Multi-family or high density high-rise housing is not the housing of last resort for households unable to afford a single-family house. Condominiums, for instance, are often the most sought after and highly appreciating real estate in many urban markets. The luxury segment of the apartment market is also rapidly expanding. Most people are surprised to learn that 41 percent of renters say they rent by choice and not out of necessity. Multifamily housing throughout the world has historically been the housing of choice by the wealthiest individuals because of the access and convenience it provides as higher-density housing has been prized for the amenity-rich lifestyle it can provide, low incidence of crime is one of the spatial quality entity the highly encouraged to dwell in the condominium or apartment high density high-rise housings.

3.6.1. Higher Density and Crime reduction

The analysis result found that when police data are analyzed per unit, condominium and apartments actually create less demand for police services than a comparable number of low density single-family houses. In Yeka-Ayat, Addis Ababa, a random sample of 400 calls for service showed that 45 percent originated from single-family houses and just 15 percent came from apartments. Similarly, a random sample of 300 calls for service in Gerji-Sunshine, Addis Ababa, found that a condominium unit's demand for police services was less than half of the demand created by a single-family house as illustrated in the *(figure 3.15)*.

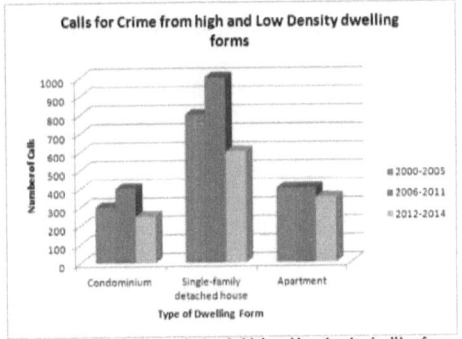

Fig. 3.15 : Rate of incidence of crime in high and low density dwelling forms in Addis Ababa, from 2000-2014 from Police Records, 2014

One reason for the misperception that crime and density are related could be that crime reports tend to characterize multifamily properties as a single "house" and may record every visit to an apartment community as happening at a single house. But a multifamily property with 150 units is more accurately defined as 150 houses. To truly compare crime rates between multifamily properties and single-family houses, the officer would have to count each household in the multifamily community as the equivalent of a separate single-family household. When they do so, many find what the previous studies prove: that crime rates between different housing typologies are comparable.

Higher-density high-rise developments can actually help reduce crime by increasing pedestrian activity and fostering a 24-hour community that puts more *"eyes on the street"* at all times. Many residents say they chose higher-density housing specifically because they felt more secure there; they feel safer because there are more people coming and going, making it more difficult for criminals to act without being discovered that is one of the physical qualities of urban spaces in built-environment. This factor shows that higher-density housing is significantly less likely to be burglarized than single-family houses. The relationships among design, management, and security became better understood in the past few decades with the publication of several seminal works, including *Defensible Space: Crime Prevention through Urban Design* by Oscar Newman and *Fixing Broken Windows: Restoring Order and Reducing Crime in our Communities* by George Kelling and Catherine Coles. Many new higher-density high-rise dwelling forms include better lighting plans and careful placement of buildings and landscaping to reduce opportunities for crime, contributing to a safer community. With the emergence of better-quality designs, higher-density high-rise mixed-use development is an attractive and safe addition to a community, one that is increasingly attracting a professional constituency seeking safety features. In fact, the luxury segment is one of the fastest-growing components of the high density multi-family industry.

3.6.2. The Role of Higher Density on Safety and Environmental quality

As it has been long argued that a well-designed higher-density high-rise community offers residents a higher-quality environment, poorly planned sprawl does the opposite. Because low-density low-rise dwelling form gobbles up so much land through large-lot zoning, it ends up destroying the very thing most people moved there for in the first place the natural areas and farmland. It forces people to drive longer distances, increasing regional *air quality problems* as aspects of environmental quality. As critically analysed in the literature, the average American man spends 81 minutes behind the wheel every day, while women average 63 minutes. And surveys show that the time spent driving has been consistently increasing every year. The national road network, currently at 1 million miles according to the Federal Department of Transportation, is still growing at an alarming rate, mainly for the purpose of connecting new low-density suburbs back to core communities. Along with the water and air pollution, construction of these highways perpetuates the cycle of sprawl, and dries up a community's financial coffers.

Increasing density in terms high-rise not only improves *air and water quality* and protects *open space* but also redirects investments to Addis Ababa City. It can revitalize existing communities and create more walkable neighborhoods with access to public transit and biking trails.

Pedestrian-friendly higher-density high-rise housing forms offer general health benefits as well. Mixed land uses give people the option to walk and bike to work, shops, restaurants, and entertainment. The convenience of compact communities may help fight diseases related to obesity. Higher-density communities are vital to preserving a healthy environment and fostering healthy lifestyles. Therefore, Higher-density high-rise comes in many forms. Some of the most attractive well-planned modern development is built at a high density high-rise with low ground coverage. Our population is changing and becoming increasingly diverse. Many of these households now prefer higher-density housing, even in suburban locations.

Although the above results can be considered subjective owing to the varying value systems different respondents attach to the different elements of a house, yet it provides some insights worth commenting upon. People seem to be most concerned and dissatisfied with the issues of privacy, noise, dust, inadequate outdoor space, and to some extent cross ventilation. Most residents in high-rise buildings expressed dissatisfaction with lack of outdoor space for their children to play. This is due to the fact that the new high-rise buildings lack courtyards and where available such courtyards have not been designed to function as outdoor spaces for living and for children play. The question of lack of cross ventilation is rather related to the observed high ground coverage's that lead to **blocked ventilation** *(see figure 3.16)*. As commented earlier, the issues of noise and dust are a result of poor road surface conditions in this part of Wube-Bereha and the many street activities that take place outside the housing environment. It is indicative therefore that high ground coverage has partly contributed to dissatisfaction among residents as far as spatial qualities are concerned.

Figure. 3.16: Low-rise housing structures-Attached dwelling units, Impossible to move through and highly vulnerable for incidences of Crime.
Dissatisfaction with security in this case has to be related with the kind of activities that take place in Wube-Bereha high density low-rise neighborhood *(see figure 3.16)*. The congested streets with street vendors and hawkers have resulted into pick-pocketing and other vices. Poor drainage is a result of increased land coverage with a poor drainage system resulting in storm water stagnation along the streets. The water supply issue is related to an overloaded system that is old and can no longer suffice the present requirements. Water rationing is frequent in Wube-Bereha and water does not reach the upper floors of the high-rise buildings due to low pressure. Despite the fact that dense environments are desirable in terms of optimal uses of land and services, residents in Wube-Bereha point to some problems of increasing density when these built environments are not properly managed.

The above (*figure 3.16*) on the other hand shows that most of the built environment is occupied by building structures or *high ground coverage* subjected to the commitment of crime. As illustrated in the above, the settlements are severely inaccessible and the situation of the security would be in question. It is strongly believed that close distance in between housing structures can directly and indirectly hinder the spatial quality of urban spaces particularly safety and security in relation to the incidence of crime. Wube-Bereha is well known with high density low-rise with high ground coverage density characteristics vulnerable for high incidence of crime and poor safety.

3.7. Results of Empirical Analysis in relation Spatial Quality

3.7.1. Statistical Model Used to evaluate the results in relation to Security

Since the study is largely an environment – behavior research, the major research instrument employed in this study were observation, household survey and field measurements. Such kinds of studies have long history. Typical studies in the past have always used behavior mapping as a way of understanding the interaction between people and space (Bechtel et al. 1987). Such an approach is premised on the fact that there is less recorded on how people use urban spaces safely and secured manner, and the kinds of dimensions and details that support different uses within such settings (Lawson, 2001; PPS - Project for Public Spaces, 2005; Goli˝cnik and Thompson, 2010). In addition to observed behavior of inhabitants living in the areas as well as the recording of size and security of the neighborhood, a questionnaire was administers to a sample of random inhabitants in four case study neighborhoods as well as measurements at plot and block levels.

This was done to comprehensively capture that active interaction between people and space. A total of 646 questionnaires were administered with the help of urban planning students in Ethiopian Civil Service University (ECSU) in 2014. Such a survey was conducted in four selected Neighborhoods from Addis Ababa including Wube-Bereha (high density-high-rise, Lideta-Firdbet (high density high-rise with low built-up area ratio/BAR/), Gerji-Sunshine (high density high-rise with high built up are ratio/BAR/) and Yeka-Ayat (low density-low-rise with extreme low built-up area ratio). Sample characteristics and a description of each study area chacteristics is summarized in table 1.

Table 3.1: Random Sampling (20% of the Total Households in each Neighborhood)

No	Case Study Neighborhood	Total Population	Total no. of households 5p=1hh	No. of households Surveyed	Households Surveyed (%)	No. of households responded	Response rate (%)
1	Wube-Bereha	6687	1337	267	20%	197	74
2	Lideta-Firdbet	6426	1285	257	20%	192	75
3	Gerji-Sunshine	6260	1252	250	20%	192	77
4	Yeka-Ayat	2070	414	82	20%	65	80
	Total	21443	4288	856		646	76.75%

Source: Author's Computations, 2014

This section statistically examines the independent variables that are considered in housing settlements when deciding better spatial quality of urban spaces to produce. The section gives a general overview of binary logistic regression, assumptions and the theoretical logistic regression model. The section also describes the expected results, obtained results and explains the results. Concludes by evaluating the obtained model.

3.7.1.1. Model results

The research hypothesis says that *""Urban Space located in a high density high-rise with low ground coverage built environment will have better spatial quality than an urban space located in high density high-rise and low-rise with high ground coverage as well as a low density low-rise with extreme low ground coverage built environment in the residential neighborhoods"*. To analyse this hypothesis a binary logistic regression was used. Binomial (or binary) logistic regression is a form of regression that is used when the dependent variable is dichotomous and the independents are continuous variables, categorical variables, or both. Binary logistic regression is used to perform logistic regression on a binary response variable (Hosmer and Lemeshow, 1989). A binary variable only has two possible values, such as *'satisfaction' (1)* on spatial quality of urban spaces or *'dissatisfaction' (0)* of spatial quality. Logistic regression applies maximum likelihood estimation after transforming the dependent into a logit variable (the natural log of the *odds* of the dependent occurring or not). In this way, logistic regression estimates the probability of a certain event occurring. Note that logistic regression calculates changes in the log of the *odds* of the dependent, not changes in the dependent itself as OLS (ordinary least square) regression does.

3.7.1.2. Assumptions of the logistic regression

Logistic regression is popular in part because it enables the researcher to overcome many of the restrictive assumptions of OLS regression (Agresti, 1990). The logistic regression handles non-linear effects even when exponential and polynomial terms are not explicitly added as additional independents because the *logit link* function on the left-hand side of the logistic regression equation is non-linear. The dependent variable need not be normally distributed (but does assume its distribution is within the range of the exponential family of distributions, such as normal, Poisson, binomial, gamma). Logistic regression does not require that the independents be interval or the independents be unbounded (Kleczka, 1980). Logistic regression assumes that error terms are *independent and all relevant variables are included in the regression model*. The logistic regression assumes a linear relationship between the logit of the independents and the dependent. Logistic regression uses maximum likelihood estimation (MLE) rather than ordinary least squares (OLS) to derive parameters (Hair et al, 1992).

3.7.1.3. Appropriateness of logistic regression model

Logistic regression can be used whenever an individual is to be classified into one of two populations. Binomial logistic regression uses a dichotomous dependent variable, which is appropriate in this case because the aim is to distinguish between two groups of residents (those who are feeling *'satisfaction'* or *'dissatisfaction'* in relation to spatial quality of urban spaces). Backward and forward variable selection procedures were used *to identify the important variables affecting the spatial quality* of the housing settlements

3.7.1.4. Theoretical logistic model

The relationship between the predictor and response variables is not a linear function in logistic regression; instead, the logistic regression function is used, which is the *logit transformation of the odds ratio Θ:*

Where, Θ = is the dependent variable (i. e. probability that an inhabitants chosen at random is agrees that the Surrounding *urban spaces spatial quality as being 'satisfied=1' or 'dissatisfied=0'*.

$$\theta = \frac{e^{(\alpha+\beta_1 x_1+\beta_2 x_2+...+\beta_i x_i)}}{1 + e^{(\alpha+\beta_1 x_1+\beta_2 x_2+...+\beta_i x_i)}}$$

Where α = the constant of the equation

β = the coefficient of the predictor variables.

x = are the explanatory variables and log is the natural logarithm

An alternative form of the logistic regression equation is:

$$\text{logit}[\theta(x)] = \log\left[\frac{\theta(x)}{1-\theta(x)}\right] = \alpha + \beta_1 x_1 + \beta_2 x_2 + ... + \beta_i x_i$$

The goal of logistic regression is to correctly predict the category of an outcome for individual cases using the most parsimonious model (Agresti, 1990). Backward stepwise regression was used. Backward stepwise analysis begins with a full or saturated model and variables are eliminated from the model in an iterative process (Afifi and Clark, 1990). The fit of the model is tested after the elimination of each variable to ensure that the model still adequately fits the data. When no more variables can be eliminated from the model, the analysis has been completed. According to Hosmer and Lemeshow (1989), there are two main uses of logistic regression. The first is the *prediction of group membership* and *provides knowledge of the relationships and strengths among the variables*. Secondly, since logistic regression calculates the *probability of success over the probability of failure*, the results of the analysis are in the form of *odds ratios*.

Therefore, the *binary logistic regression model* was applied to determine factors that explained why some urban citizens define their surroundings spatial quality of urban spaces in terms of *satisfaction* (1) and *dissatisfaction* (0) (satisfied and dissatisfied). When dealing with a dichotomous dependent variable (response variable) - the main interest is to assess the probability that one or the other characteristic is present (Peng and So 2002; Peng et al, 2002). The logistic regression model answers the question what determines the probability that the answer is *satisfied, or dissatisfied*. The special features of the model guarantees that probabilities estimated from the logistic model will always lie within the logical bounds of *0 and 1*. In other words the probability that an urban citizen picked at random is defined *spatial quality of urban spaces* in the housing settlement as *spatially satisfied or dissatisfied*, quality is not a continuous variable but a discrete one. The logistic regression model can be expressed mathematically as shown above;

The selection of predictor variables was based on the review literature, observation, standards and norms on *'spatial quality'* and *'density'* along with patterns of built-form issues.

The *performance of good spatial quality* can be predicted based on: High density high-rise, low ground coverage, high total floor area ratio, outdoor spaces, green and open spaces, accessibility, safety and security, plot exposure, ventilation and interactions. Spatial quality is measured in dichotomous scale *(satisfied=1 and dissatisfied=0)* and it has 8 independent variables identied that fit the model. Some predictor variables were avoided because of the redundancy of data records like *Plot exposure vs. Ventilation, High density-high-rise vs dwelling/population density,* and etc. one of them has been avoided due to overlapping of data records and may have not bring significant change in the model.

3.7.1.5. Expected results

The aim of performing the logistic analysis was to determine those factors that influence spatial quality of urban spaces in the housing settlements. The variables (x's) that are expected to influence spatial quality of urban spaces whether leads to ***satisfaction or dissatisfaction***, include:

a) **High density high-rise dwelling forms**: Multi-family multi-storied dwelling forms influences the spatial quality of urban spaces in the built-environment by creating better possibilities for more and *more pedestrian movements* to make vibrant spaces, which would combat incidence of crime. It also minimizes the employement of guards in the neighbourhoods, creating possibilities for 'Many eyes on the street' = Natural surveillances in instead of employing guards. It has better access for communal amenities and services. These can positively influence spatial quality of housing settlement. It also offers better space and utility efficiency than low density low-rise dwelling forms.

b) **Low ground Coverage:** the land coverage without built-up density thresholds influences built-environments in the provision process of outdoor spaces, green and open spaces, ventilation and amenities. The higher the ground coverage above the recommended threshold of Built up area ratio /BAR/ blocks probabilities of having better quality urban spaces. Therefore, Low BAR is one of the predictory variables which affect spatial quality of urban spaces. High BAR blocks spaces for amenities, parking, outdoor spaces, spaces for children to play etc.

c) **High-FAR:** it is also one of the density characteristics which influence built-environment in building quality urban spaces. Together with above *Low BAR* density characteristics, it plays significant role in creating better possibilities for active pedestrian movements in the neighborhoods in combating against crime, adequate outdoor spaces, amenities, parking spaces. Therefore, it is a kind of density characteristics which can also determine spatial quality of urban spaces.

d) **Set-Backs:** There are four types of set-backs including front, rear, and side and neighboring set-backs; however this study focuses with *front set-back* that has direct relationship with main streets that the buildings are facing. Set-back is important characteristics of plot or building that would highly support circulation and ventilation as well as street side greenery and opens spaces in the built environment. So that it has its own contribution for better spatial quality of urban spaces.

e) **Plot Exposure:** Plot exposure is one of the plot characteristics which influences the built-form patterns and spatial quality by allowing cross ventilation and spaces for amenities and public spaces like streets and greeneries. The lesser the exposure the lesser the ventilation and access to amenities and public spaces like streets etc. Therefore, plot exposure is found to be the determinants of spatial quality. The better the plot exposures the better spatial quality of urban spaces are expected. The plot or plot exposure less than 2 leads to poor ventilation and access to amenities leads to dissatisfaction of of residents with respect to spatial quality of urban spaces

f) **Outdoor space:** It is one the predictory variables which is expected to influence spatial quality by offering spaces for children to play, families to recreate, amenities, parking spaces, greeneries around or in front or rear or at the sides of the building. It is also directly related with Low BAR. If the plot or block is going to be occupied dominantly by building structures, the possibility of having outdoor spaces is going to be blocked and leads to dissatisfaction of residents with regard to spatial quality might be occurred. So outdoor spaces is one of the predictory variables which influence spatial quality.

g) **Green and open spaces**: they are also other determinant factors, which affect spatial quality in the housing settlements. Green and open spaces are very important to keep urban sustainability all over the world, especially cities in hot dry and semi humid cities of the world like Addis Ababa. They are considered as extension of home in such harsh temperature, to maintain microclimate balance. So, green and open spaces would influence spatial quality directly.

h) **Accessibility and Circulation:** They are also very important predictory variables which affect spatial quality of urban spaces. Most of the neighborhoods in Addis Ababa are inaccessible and subjected to very limited circulation spaces. So neighborhoods with limited alleys and foot paths, no standards streets within neighborhoods or no accessibility and mobility leads to poor spatial quality. Therefore, accessibility and circulation are expected to influence and change spatial quality. In some locations a significant portion of these accessibility needs can be met by walking. **Freedom of movement** and access to certain activities and destinations are recognized in many cultural and legal traditions of the city. But in many neighborhoods of Addis Ababa freedom of movement is limited due to absence of standard streets as it could be taken as indicator of poor spatial qaulity of urban spaces.

i) **Security:** Security is an important attribute to determine which spatial quality is likely to satisfy the residents in the housing settlements. Various categories of crimes in the neighborhoods including petty crimes to killings, burglary, vandalisms, thefts, assault etc are the main challenges in the built environments of Addis Ababa. So security is one of the determinant factors which influence spatial quality of urban spaces in the settlements. This predictory variable is highly correlated with high density high-rise dwelling forms to combat against crimes by forming *"many eyes on the street"* for natural surveillances as stated above. Because high density high-rise creates probabilities of more pedestrian movements to combat crimes before being dicovered.

It is assumed that ***spatial quality of urban spaces*** in the housing settlement, this will be the dependent variable (criterion value), which is the object of classification efforts.

3.7.2. Model Evaluation with respect to Null and Alternative Hypothesis

Parameters in logistic regression model were estimated using the maximum likelihood method. The statistical significance of each coefficient was evaluated using the Wald test. In this analysis, the enumerated regression coefficients represent the change in the ***logit of the probability*** from a unit change in the associated predictor, assuming other factors are constant (Gujrati, 2003).

The goodness-of-fit test of the regression model in this study was analyzed using;

1. The *Omnibus test*, which is a likelihood ratio chi-square test that test whether the coefficients of the variables in the model are all jointly equal to zero.
2. The Hosmer & Lemeshaw (H-L) goodness-of-fit test, which examines the null hypothesis that the model adjust well to the data and
3. The Cox and Snell (1989) and Nagelkerke (1991) – two descriptor measures that reveal the amount of variation in the outcome variable that is explained by the models (Long, 1997; Hosmer and Lemeshow, 2000).

Omnibus Tests of Model Coefficients

		Chi-square	df	Sig.
Step 1	Step	404.367	8	.000
	Block	404.367	8	.000
	Model	404.367	8	.000

Model Summary

Step	-2 Log likelihood	Cox & Snell R Square	Nagelkerke R Square
1	452.141ᵃ	.465	.633

a. Estimation terminated at iteration number 6 because parameter estimates changed by less than .001.

Hosmer and Lemeshow Test

Step	Chi-square	df	Sig.
1	10.904	8	.207

The Hosmer & Lemeshow (H-L) inferential goodness-of-fit test yielded a Chi-square (8 degrees of freedom) of 10.904 and was insignificant (p > 0.05, which is equivalent to 0.207>0.05) suggesting that the model fitted to the data well and clearly indicates the **good potential predictor variables** for the Model. Two other descriptive measures of goodness of fit are R^2 indices defined by Cox and Snell (1989) and Nagelkerke (1991). Results suggest 35.4% to 56.2% of variations in the outcome (i.e. the probability of a randomly chosen urban residents agreeing that spatial quality in the housing settlement is good and showing the feeling of satisfaction. On the other hand the predicted percentage by the model is **85.3%, which is above 70% and is very good model** for the future improvement of spatial quality of urban spaces in the city

Table 3.2: Test parameters for the binary logistic model

Variables not in the Equation

			Score	df	Sig.
Step 0	Variables	Highdensity_Highrise	200.876	1	.000
		Low_groundcoverage	128.676	1	.000
		Plot_exposure	110.844	1	.000
		High_FAR	106.190	1	.000
		Outdoor_Space	88.345	1	.000
		Green_Openspaces	37.505	1	.000
		Security	63.471	1	.000
		Accessibility	19.720	1	.000
	Overall Statistics		311.610	8	.000

Classification Tableᵃ

		Predicted		
		Spatial_Quality		Percentage Correct
Observed		.00	1.00	
Step 1	Spatial_Quality .00	359	43	89.3
	1.00	52	192	78.7
	Overall Percentage			85.3

a. The cut value is .500

The p-value for each term tests the null hypothesis that the coefficient is equal to zero (no effect). A low p-value (< 0.05) indicates that Author can reject the *null hypothesis*. In other words, a predictor variable that has a low p-value is likely to be a meaningful addition to Author's model because changes in the predictor's value are related to changes in the response variable (*See table 3.3 below*).

$H_0=0$, the author can reject null hypothesis.

- Wald statistic is the significance test for each parameter in the model
- Null is that each $\beta = 0$

3.7.3. Results of Alternative Hypothesis Testing

Results reveal that out of the 646 surveyed household heads of four sampled neighborhoods in Addis Ababa, *a total of 62.2 % believed to be responded that spatial quality* of housing settlement is poor/dissatisfied/. A number of constraining factors were discerned. Such constraints were analysed using the **binary logistic regression model**. Results are summarized in *table 3.3 below*. Backward stepwise procedure was used to identify the important variables affecting the spatial quality of urban spaces. No variables removed from the model because all the variables affect spatial quality significantly except security. Even Security itself affects it in certain extent as shown *below in the table 3.3*, so that security also maintained in the model. All variables are statistically significant at <0.05. The p values are as shown in the *table 3.3 below*.. All the predictory variables represent positive betas, which could be taken as good news.

Classification Table[a,b]

		Predicted		
		Spatial_Quality		Percentage Correct
Observed		.00	1.00	
Step 0	Spatial_Quality .00	402	0	100.0
	1.00	244	0	.0
	Overall Percentage			62.2

a. Constant is included in the model.
b. The cut value is .500

Table 3.3: Test parameters for the binary logistic model

Variables in the Equation

		B	S.E.	Wald	df	Sig.	Exp(B)
Step 1[a]	Highdensity_Highrise	.284	.033	72.285	1	.000	1.328
	Low_groundcoverage	.142	.035	16.478	1	.000	1.153
	Plot_exposure	.108	.027	16.198	1	.000	1.114
	High_FAR	.161	.034	21.756	1	.000	1.174
	Outdoor_Space	.164	.038	18.890	1	.000	1.178
	Green_Openspaces	.104	.042	6.274	1	.012	1.110
	Security	.040	.042	.908	1	.341	1.040
	Accessibility	.074	.030	5.833	1	.016	1.076
	Constant	-5.882	.497	140.187	1	.000	.003

a. Variable(s) entered on step 1: Highdensity_Highrise, Low_groundcoverage, Plot_exposure, High_FAR, Outdoor_Space, Green_Openspaces, Security, Accessibility.

The positive beta estimate on high density high rise, low BAR/low ground coverage, and high total floor area ratio, accessibility, outdoor spaces, green and open spaces imply that spatial quality of housing settlement is influenced by high density high-rise 1.328 times (ie. **Exp β = 1.328**), low BAR 1.153 times, and high total floor area ratio 1.174 times more likely to increase the spatial quality of urban spaces in the area than low density low-rise dwelling forms, and with high FAR or high floor area ratio 1.174 times better than low FAR, neighborhood quality in the built-environment. Spatial quality was also found to vary significantly with the Greenery and open spaces ($p < 0.05$). It is very important to note that *"the negative beta estimate reveals that spatial quality of urban spaces were dismissed as highly dissatisfied (in relative terms) as far as spatial quality of housing settlement is concerned"*, but in this particular case no negative beta was observed from the logistic regression analysis results. The model also indicates that indiscriminate application of density including Law BAR and High FAR influence the quality of urban spaces. Therefore, every unit increase in *'X'* or **predictory variable** has an exponential effect on the odds of success so an *'odds ratio'* can be >1. All predictory variables have been shown positive relationship with spatial quality of urban spaces.

So, The Wald statistic and the corresponding significance level test, the significance of each of the covariate and dummy independent variables in the model are shown in the above table. If the Wald statistic is significant (i.e., less than 0.05) then the parameter is significant in the model. Of the independent variables, security is insignificant, whereas *high-density high-rise, plot exposure, Low BAR, High FAR, Outdoor space, Green and open space and accessibility* have significantly affected the spatial quality of urban spaces in the housing settlements. Even, security itself has 1.04 *'odds ratio'* that means, it has **1.04** better performances than low-rise with high ground coverage *(see table 3.3)*. The model clearly unveils that increasing or changing one unit of predictory variables changes the effects of the other, and decreasing also would affect the significance of the other. Therefore, increasing the values of explanatory variables changes the probability of being improved. The Model equation is as Follows:

Box 10.3: Derived Equation

(Spatial Quality) = Θ = -5.882 +.284highDensity high-rise +.142low-BAR + .108Plot_exposure + .161floor area ratio + 0.164outdoor space + 0.104green & open space .04Security +0.074Accessibility

$\Theta = \ln(P/1-P) = -5882 + .284X_1 + .142X_2 + .108X_3 + .161X_4 + .164X_5 + .104X_6 + .04X_7 + .074X_8$

$$\Theta = \ln\left(\frac{P}{1-P}\right) = \frac{e^{-5882 + .284X_1 + .142X_2 + .108X_3 + .161X_4 + .164X_5 + .104X_6 + .04X_7 + .074X_8}}{1 + e^{-5882 + .284X_1 + .142X_2 + .108X_3 + .161X_4 + .164X_5 + .104X_6 + .04X_7 + .074X_8}}$$

Θ = Probability of an inhabitants agrees that the spatial quality of urban spaces are good spatially in the Neighborhood, also could be taken Spatial quality (i.e. probability / *Satisfy = 1, Dissatisfy=0*)

X_1 = High Density High-rise (10=> 5 storeys and >625inh/ha; 8 = 3-4 storeys & 400-625inh/ha; 5= 2-3 storeys 280-400inh/ha; 2= 1-2 storeys 250-280inh/ha; 0= 1 storeys <250inh/ha)
X_2 = Plot Exposure (10= 3-4; 8= 2-3; 5= 1-2, 2 = 1; 0=0)
X_3 = Ground Coverage Ratio or BAR/ (10=0.30-0.40; 8=0.40-0.50; 5= 0.50-65; 2= 0.65-0.75; 0=>0.75)
X_4 = Floor area Ratio /FAR/ (10 = >2.5; 8=2.0-2.5; 5 = 1.5-2.0; 2 =1.15-1.5; 0=<1.15)
X_5 = outdoor-space (10= strongly agree; 8= agree; 5= Neutral; 2= Disagree; 0 = Strongly Disagree)
X_6 = Greenery-open (10= strongly agree; 8= agree; 5= Neutral; 2= Disagree; 0 = Strongly Disagree)
X_7 = Security (10= strongly agree; 8= agree; 5= Neutral; 2= Disagree; 0 = Strongly Disagree)
X_8 = Accessibility (10= strongly agree; 8= agree; 5= Neutral; 2= Disagree; 0 = Strongly Disagree)

In this binary regression model, it is possible to test the significance of each pattern of built-forms on the basis of the above logistic regression analysis, overall results.

3.7.3.1. High density high-rise with low ground coverage by buildings

As it has been long discussed that *Lideta-Firdbet* was one of the four case study settlements characterised by *low BAR vs. high FAR* or *high density high-rise with low ground coverage by buildings*. From the above analysis result, the major identified explanatory variables, which affect spatial quality of urban spaces were density (high or low, built-up density), dwelling forms, plot characteristics including (exposure, size, plot ratio and plot boundary), green and open spaces, communal outdoor spaces, interactions, safety and security, accessibility and circulation, ventilation and daylight access. The above logistic regression model has not been analysed all the predictory variables against response or dependent variable (spatial quality). Some independent variables might have similar predictory data records, and hence the author has been decided to accept one of them. So the number of predictory variables may decrease in number due to redundancy of similar values of data record. For instance, the data record for plot exposure and ventilation, high density high-rise and population density and etc are almost the same. So the author was forced to avoid one of them due to overlapping, because it highly threats the reliability and fitness of the model while testing the hypothesis of the study.

Therefore, the logistic regression analysis result unveils that the performances of high density high-rise are 1.328 times higher than that of the performances of low density low-rise in relation to spatial quality of housing settlement is concerned. It is also crystal clear that high density high-rise and low BAR vs. high FAR are similar in patterns of built-forms. So that this built-form pattern performs 1.328 times than high BAR and low FAR. As far as total floor area ratio is concerned, 1.153 times better performance of high FAR is observed, than that of the Low FAR in the built-environments. The performance of plot exposure is 1.114 times better than that of the performances of low exposures in built-environment as far as spatial quality of housing settlement is concerned. The above output table significance (P<0.05) of *high density high rise, Low BAR, High Floor area ratio, outdoor space, green and open spaces* and accessibility were influenced the spatial quality of urban spaces. Actually, from the above *Table 10.2*, it is observed that, all the variables*(high density high rise, plot exposure, low BAR, High FAR, outdoor space, greenery and open spaces and accessibility)* except category of *security* are found to be significant. Even security itself has a positive β and in certain extent it influences as a unit change in security may change other predictory variables. Therefore, the author can easily interpret a regression coefficient; say $β_1$, as expected change in log of *"θ"* with respect to a one unit increase in "X_1" holding all other variables at any fixed values, assuming that 'X_1' enters the model only as a main effect.

This clearly unveils that low BAR vs. High FAR built-form would offer better quality urban spaces in the housing settlement. The changes in Low BAR can change the other predictory variables in the housing settlements including high density high-rise, plot exposure, high Floor area ratio, accessibility, security, outdoor spaces, and greenery and open spaces etc. Similarly high density high rise has direct impact on the spatial quality of housing settlements. Actually high density high-rise settlement is highly correlated with high floor area ratio in terms of population density. Therefore, the changes in ground coverage (BAR) and high floor area ratio might have directly change outdoor spaces, greenery, density and the likes in the built environment as illustrated by this Model. In the same manner the changes in high density high rise built environment can change the other predictory variables as spatial quality of the housing settlement is concerned. And hence spatial quality of housing settlement is strongly correlated with all the above explanatory or predictory variables.

It is also worth mentioning that this statistical model indicates high density high-rise with low ground coverage has posive impact on the spatial quality of housing setlement by offering spaces for children to play, adequate amenities, greenery and open spaces and so on. Thus, high floor area ratio with low ground coverage would create possibilities for greenery and open spaces, ventilation, space for walking, standing, sitting and protection. Protection particularly through the provision of trees for microclimate balance, shading effect as well securing safety and security. Security is supposed to minimize incidence of crime by natural surveillance through high density high-rsie building by installing *'many eyes on the street'* from the residents of high rise buildings.

It is very important to retrace research question and objectives underlined that at what extent density and spatial qualities are correlated as well as what impacts density exert on spatial qualities of housing settlement. Accordingly the logistic regression analysis has clearly been illustrated the relationship between spatial quality and Density in terms of *low ground coverage, high total floor area ration and high density high-rise* which are the significant explanatory variables that might affect spatial quality of housing settlements as tested in Wald test.

3.7.3.2. High density Low-rise with High ground coverage by buildings

Wube-Bereha is one of the case study settlement characterised by *high density low-rise with extreme high ground coverage or high BAR vs. low FAR built-form patterns* as well as overcrowdness and congestion. As far as high density low-rise with high ground coverage by buildings is concerned, the above logistic regression analysis result is found to be the opposite, as the performances of high density low-rise are 1.328 times lower than that of the performances of high density high-rise in relation to spatial quality of housing settlement is concerned. It is evident that high density low-rise and high BAR vs. low FAR are similar in patterns of built-forms. And hence this built-form pattern performs 1.328 times lower than low BAR vs. high FAR. As far as total floor area ratio is concerned, 1.153 times worse performance of low FAR is observed, than that of the high FAR in the housing settlement. The performances of plot exposure are 1.114 times lesser than that of the performances of high exposures in built-environment as far as ventilation is concerned as predictory variable of spatial quality of housing settlement.

Similarly, *high density high rise, Low BAR, High Floor area ratio, outdoor space, green and open spaces* and accessibility were highly influenced the built environment in terms of spatial quality. Particularly, these predictory variables are highly limited in high density low-rise with high ground coverage pattern of built-forms. For instance, the land is dominantly occupied by housing structures in this neighborhood, so that it is composed of extreme challenges of *exposure, green and open space, outdoor spaces, accessibility, circulation and security*. The performance of low BAR is 1.174 times likely to offer better spatial quality than that of high BAR as far as the type of built-form pattern is concerned. Actually, from the above *Table 3.3*, it is observed that, all the variables(*high density high rise, plot exposure, low BAR, High FAR, outdoor space, greenery and open spaces and accessibility*) except category of *security* are found to be significant for the built-form type of *low BAR vs high FAR*. So the opposite is true for *High BAR vs. low FAR*. Therefore, we can easily interpret a regression coefficient; say $β_I$, as expected change in log of "$θ$" with respect to a one unit increase in "X_1" holding all other variables at any fixed values, assuming that X_1 enters the model only as a main effect. So changes in each of the predictor variables might have change the other variables. Therfore, the model clearly unveiled that high density high rise with low ground coverage built-form would offer better quality urban spaces that high density low-rise with high ground coverage built-form patterns as concluded below.

CHAPTER 4: KEY FINDINGS & CONCLUSION

3.1. Key Findings

- The study found that density is a critical typology and integral component of urban planning in determining sustainable *patterns of built-forms and quality of urban spaces*. The relationship between density and urban character is also based on the concept of *viable thresholds (BAR: 30-50% on the basis of floor area ratio/FAR<2.5/)*: at certain *densities (thresholds) (>625 inh/ha)*, the number of people within a given area becomes sufficient to generate the interactions needed to make urban functions or activities viable. Therefore, this study concludes in a wider sense, sustainable cities are a *matter of density*. Density should be used as **prescriptive** and norms to design built-environment rather than **describing** built-environment.

- The study was identified *Plot characteristics, Building Characteristics, Density Characteristics* as the determinants of patterns of built-forms. With this regard, the study found Wube-Bereha as high density low-rise with high ground coverage that most of the plots have not exposures as plot characteristics facing high limitation of ventilation and daylight access apart from other challenges of spatial quality elements.

- Similarly, Absence of **Proper and optimum set-backs** (>3m) has been highly influenced the patterns of built-forms while designing and planning built-environment so as to promote the spatial quality of urban spaces in terms of creating possibility to allow adequate ventilation and circulation as well as street side greenery and gardening and etc. Absence of Setback has also been highly influenced the percentage of ground coverage in the built-environment and directly affected the aesthetics of the neighborhood as well. So, particularly front setback has contribution for strong relationship between built-up density and spatial quality of urban spaces.

- The study also found that *Patterns of Built-Forms* are the *outcomes or product of density* that would influence spatial quality. *High ground coverage* versus high and low floor area ratio is the basic density characteristics of developing countries cities that leads to limited possibility of providing outdoor spaces, less access to mobility and circulation, create possibilities for incidence of crime, less opportunity to green and public urban spaces in the city. And hence this pattern of the built-form highly threatened the spatial quality of urban spaces. The neighborhoods could also have different patterns of built-forms with constant population or dwelling density including high density high-rise with low ground coverage, medium density medium rise with medium coverage and high density low-rise with high ground coverage. Each built-form patterns has different characteristics and implications.

- *The other essential finding is high density low rise with high ground coverage and high density high rise with high ground coverage* are the patterns of built forms and density characteristics, which have been highly threatened the spatial quality elements by limiting possibilities to provide adequate green and open spaces, Ventilation, Circulation and mobility, safety and security, Because most of the spaces were occupied by housing structures and close distance between housing structures. The study found that *high density high rise with Low ground Coverage built-forms* has been offered higher possibility for building better spatial quality elements in the built environments.

- This study also found that *"Start Planning and Design of Built Environment with proper density thresholds"* could be taken as the best strategy in the process of planning and designing of built environment so as to install good spatial quality of urban spaces. Thus, urban spaces design in the housing settlements should be considered as critical policy element in Addis Ababa city administration and in the urban areas of the country as the whole. *"Start Planning and Design of built environment, which are supposed for housing settlement with prescription of 'density thresholds' (BAR: 30-50%, FAR>2.5, Po^{pn} Density >625)) to offer quality urban spaces in the city"*. Different Built-form patterns and their implications in relation to spatial quality have been summarised & the four case study areas analysis results and their implications clearly indicated as:
  - ✓ *Lideta-Firdbet* was characterized by **Low BAR(<45%) vs. High FAR(>2)** Built-Form pattern showing better quality of urban spaces in the settlement as explicitly addressed, **High Density high-rise** housing settlement with *low ground coverage*,
  - ✓ *Wube-Bereha* was characterized by **high BAR (85-110%) vs. Low FAR (<1.15)** built-form pattern and High Density low-rise with high ground coverage composed of poor spatial quality urban spaces in the built environment.
  - ✓ *Gerji-Sunshine* was characterized by **High BAR (80-85%) vs. High FAR (>2) built-form pattern and high density high rise** density category offering low possibilities to install spatial quality elements of urban spaces. Because much spaces has been occupied by building structures and
  - ✓ *Yeka-Ayat* was characterized by **Low BAR (<30%) vs. Low FAR(<1)** built-form pattern and **Low Density Low-rise** built-environment has shown better quality spatially than Wube-Bereha high density low-rise, but high insecurity challenges has been recorded in Yeka-Ayat.
- On the other hand, the study found that Population density has been received considerable attention as it relates to crime. The analysis result unveiled that the relationships between density and crime, found positively correlated relationships between crime and density, particularly high density high-rise dwelling forms with low ground coverage by housing structures. Thus *high density high-rise dwelling forms* have lower commitments of crime than low density low-rise detached dwelling units in the built environment.
- The othe key finding of this study is that *high density high-rise* dwelling forms are believed to combat incidence of crime through *'natural surveillance'* by creating *'many eyes on the street'* like multi-family Condominiums and apartments in the urban environments. And hence, high density high-rise dwelling form with low ground coverage can better reduce the incidence of crime than low density low-rise dwelling forms, because high density high-rise with low ground coverage built-environment could have a possibility to offer adequate public amenities, communal outdoor spaces for people to interact, inhabitant to relax in green/open spaces, such circumstances on the other hand would create *24-hours active pedestrian movement* in the area. It is also worth mentioning that active pedestrian movement and residents in the high density high-rise housing forms would create less probabilities of incidence of crime in the settlement due the formation of *'Many Eyes on the street'* through Natural Surveillances.

3.2. CONCLUSION

On the basis the literature reviewed and the emperical analysis results, the study concludes that *density* is a critical typology and integral component of urban planning in determining sustainable *the patterns of built-forms and spatial quality of urban spaces*. Therefore, the relationship between density and urban character is also based on the concept of *viable thresholds (BAR: 30-50% on the basis of floor area ratio/FAR<2.5/)*: at certain *densities (thresholds) (>625 inh/ha)*, the number of people within a given area becomes sufficient to generate the interactions needed to make urban functions or activities viable. And hence, this study concludes in a wider sense, sustainable cities are a matter of density. *Density should be used as **prescriptive** and norms to design built-environment rather than **describing** built-environment.*

The study also concludes that *Patterns of Built-Forms* are the *outcomes or product of Density* in the built-environment that would influence spatial quality of housing settlements. *High ground coverage* versus high and low floor area ratio is the basic density characteristics of developing countries cities that leads to less possibility of outdoor spaces, less access to mobility and circulation spaces, create possibilities for incidence of crime, less opportunity to green and public urban spaces in the city. And hence this pattern of the built-form highly threatened the spatial quality of urban spaces. It is also possible to conclude that *high density low rise with high ground coverage and high density high rise with high ground coverage* are the patterns of built forms and density characteristics, which have been highly threatened the spatial quality elements, because most of the built-environment is occupied by building structures. *High Density high rise with Low ground Coverage built-forms* has offered high possibility for building higher spatial quality elements like space for circulation, green and open spaces, ventilation and air circulation within building structures, security due to many eyes on the street and open spaces around the buildings, access day light, possibility to use private and communal outdoor spaces.

Therefore, on the basis of the analysis results and review from the preceding sections, the Author can conclude that little attention has been given to *density and patterns of built-forms* as basic prerequisites of sustainability of urban built-environment as well as integral part urban planning and design. This study concluded that *"Start Planning and Design of Built Environment with proper density thresholds"* that could be underlined as the best strategy in the process of planning and designing so as to build appropriate patterns of built-form to make sure sustainable quality of urban spaces. Thus, urban spaces design in the housing settlements should be considered as critical policy element in Addis Ababa city administration and in the urban areas of the country as the whole.

Finally, recent debates about the creation of more sustainable urban built-form, compact cities have led to a renewed focus on issues of density, especially dwelling & built-up density. The argument is that compact cities form or high density *urban built form* can offer a high quality of life in the built environment through un-indiscriminate application of *'density'* by installing better spatial quality elements in the city. The study concluded that density is an important factor to consider in designing safer environments. It is also worth mentioning that Multi-family or *high density high-rise dwelling form* is not the housing of last resort for households unable to afford a single-family house. Condominiums, for instance, are often the most sought after and highly appreciating real estate in many urban markets. Multifamily housing throughout the world has historically been the housing of choice by the wealthiest individuals because of the access and convenience it provides as higher-density housing has been prized for the amenity-rich lifestyle it can provide, low incidence of crime is one of the spatial quality entity the highly encouraged to dwell in the condominium or apartment high density high-rise housings.

Furthermore, the study concludes that *higher density high-rise with low ground coverage generates more flows and movements, which provide natural surveillance on the streets.* Generally, much depends on design and other factors, increased intensity of human activity and 24-hour use of public spaces can promote safer urban environments through *"many eyes on the street"* and more economically dynamic environments. Higher residential population densities can, if appropriately configured, create a *"critical mass"* for pedestrian access to parks, communal outdoor space to people to socialize/interact each other, community facilities. Therefore, the analysis result unveiled that high-density high-rise with low ground coverage offers opportunity and possibility to combat the incidence of crime as well as to install safety in residential neighborhoods.

REFRENCES

- Addis Ababa structure plan, (2011). *Building height component-revised*, Addis Ababa, Ethiopia.
- Alexander E., (1993), Density Measures: A Review and Analysis, *Journal of Architectural and Planning Research*, Vol. No. 10, No.3, pp.181-202.
- Acioly C. and Davidson F. (1996). *Density in Urban Development*, Building issues,Vol.8, No.3, Lund Centre for Habitat Studies, Lund University.
- Bechtel, S.L., Elman, B.R., & Jordon, J.L. (1984). *Skier's knee: "The enunciate connection"*, The Physician and Sports medicine, 12, 5 1-54.
- Bright, J. (1992). *Crime Prevention in America*. Chicago: University of Illinois at Chicago
- Ceccato V. A., (2001), Understanding Urban Patterns, Qualitative and Quantitative Approaches, Department of Infrastructure and Planning, Royal Institute of Technology, Stockholm, Sweden C. Rowe, Koetter, (1978). *The Collage City*, New York.
- Carmona, M., Heath, T., Oc, T. And Tiesdell, S. (2003), *Public Places–Urban Spaces*. The Dimensions of Urban Design, London: Architectural Press.
- Cheng, V. (2010). *The Understanding of High Density*. En E. Ng, *Designing high-Density cities for social and environmental sustainability*. London: Earthscan.
- C. Ray Jeffery (1971), *Crime Prevention through Environmental Design*, New York, USA
- Frank, L. and Pivo, G. (1994). *Impacts of mixed use and density on utilization of three modes of travel*: single-occupant vehicle, transit, and walking, pp44-52.
- Frey, H. (1 999), *Designing the City*, Towards a More Sustainable Urban Form, London.
- Gehl, Jan, (2010). *Cities for People,* Island press, pp. 105.
- Galle, O., W. Gove, and J. M. McPherson (1972) *"Population density and social pathology*: what are the relationships for man?" Science 176 (April): 23-30.
- Gardiner, R.A. (1978). *Design for Safe Neighborhoods*: The Environmental Security Planning and Design Process. Washington: U.S. Government Printing Office
- Goličnik, B., Thompson, C.W. (2010) *Emerging relationships between design and use of urban park spaces*. Landscape and Urban Planning, 94(1), 38–53.
- Gómez Arenas A., (2002). *Analysis of Infrastructure Provision in Low-income Settlements*, Port Elizabeth South Africa, Master's Thesis, EESI Programme, Royal Institute of Technology, Stockholm.
- Jacobs, J. (1961). *The Death and Life of Great American Cities. New York*: Vintage Books/Random
- Jacobs, A. (1985). *Looking at cities*. Cambridge, Mass: Harvard University Press.

- Jacobs, Jane, (1992). *The Death and Life of Great American Cities*, Vintage Books Edition,
- Jencks, M., Burton, E. and Williams, K. (ed.), (1996). The Compact City: A Sustainable Urban Form. London:
- JO, S. (1998). *Spatial Configuration and Built Form*, In Journals of Urban Design, Vol: 3, No: 3.
- Knox, P., Pinch, S. (2000). *Urban Social Geography* – An Introduction -,England:, pp.77-126
- Kostof, S. (1991). *The City Shaped*–Urban Patterns and Meanings through History, London: Thames
- Krier, R. (1979). *Urban Space*, New York: Academy Editions.
- Lozano, E. (1990). *Density in Communities, or the most important factor in building urbanity*. En E. Lozano, *Community Design and the Culture of Cities: the Crossroad and the Wall*. Cambridge, Massachusetts: MIT
- Lynch, K. (1994). *A Theory of Good City Form*, Massachusetts: Massachusetts Institute of Technology.
- Lynch, K. (1981). *A Theory of Good City Form*. Massachusetts: MIT Press.
- Madanipour, A. (1996). *Design of Urban Space*, New York: John Wiles & Sons.
- Matthew Carmona and Louie Sieh (2004). *Measuring quality in Urban Planning, Managing the performance process*, London And New York
- Moudon, A. V. (1997). *Urban Morphology as an Emerging Interdisciplinary*, Urban Morphology 1, pp. 3.
- Newman O., (1973), Defensible Space; Crime Prevention Through Urban Design, New York.
- Oswald, F. and Bacini, P. (2003). *Designing the urban*, Basel, Boston, Berlin: Birkhäuser
- Oscar Newman, (1972). *The Defensible Space*, New York, US America
- Poyner & Webb, (1992). *Their evaluation of the changes showed great reductions in the level of thefts after the redesign efforts*
- P.H.M. Steven *(1960), Density of Housing areas,* London, UK
- Punter, J., and Carmona, M. (1997). *The Design Dimension of Planning*, London: E&Fn Spon.
- Rådberg J. (1996). *Towards a theory of sustainability and urban quality: A new Method of Typological Urban Classification* in Gray M., (ed.). *Evolving Environmental Ideals: Changing ways of life, Values and Design*
- Rapoport, A. (1977). *Human Aspects of Urban Form*, Oxford: Pergamon Press.
- Spreiregen, P. D. (1965), *Urban Design: The Architecture of Towns and Cities*, New York: McGraw-Hill.
- Trancik, R. (1986*). Finding Lost Space – Theories of Urban* Design -, New York: Van Nostrand Reinhold.
- UNCHS, (2011) Campaign for good urban governance, Nairobi, Kenya.
- Yin, R. (1994) *Case Study Research, Design and Methods,* Thousand Oaks, California: Sage.
- W. C. Sullivan, Kuo, & De Pooter, (2001*). Environment and Crime in the Inner City*: Does Vegetation Reduce Crime?
- Wilson, J. Q. and R. J. Herrnstein (1985). *Crime and Human Natur*e. New York: Simon and Schuster.

I want morebooks!

Buy your books fast and straightforward online - at one of the world's fastest growing online book stores! Environmentally sound due to Print-on-Demand technologies.

Buy your books online at
www.get-morebooks.com

Kaufen Sie Ihre Bücher schnell und unkompliziert online – auf einer der am schnellsten wachsenden Buchhandelsplattformen weltweit! Dank Print-On-Demand umwelt- und ressourcenschonend produziert.

Bücher schneller online kaufen
www.morebooks.de

OmniScriptum Marketing DEU GmbH
Heinrich-Böcking-Str. 6-8
D - 66121 Saarbrücken

Telefax: +49 681 93 81 567-9

info@omniscriptum.de
www.omniscriptum.de

www.ingramcontent.com/pod-product-compliance
Lightning Source LLC
Chambersburg PA
CBHW031541210526
45464CB00003B/1096